Aims and Scope

Urban and Landscape Perspectives is a series which aims at nurturing theoretic reflection on the city and the territory and working out and applying methods and techniques for improving our physical and social landscapes.

The main issue in the series is developed around the projectual dimension, with the objective of visualising both the city and the territory from a particular viewpoint, which singles out the territorial dimension as the city's space of communication and negotiation.

The series will face emerging problems that characterise the dynamics of city development, like the new, fresh relations between urban societies and physical space, the right to the city, urban equity, the project for the physical city as a means to reveal *civitas*, signs of new social cohesiveness, the sense of contemporary public space and the sustainability of urban development.

Concerned with advancing theories on the city, the series resolves to welcome articles that feature a pluralism of disciplinary contributions studying formal and informal practices on the project for the city and seeking conceptual and operative categories capable of understanding and facing the problems inherent in the profound transformations of contemporary urban landscapes.

Urban and Landscape Perspectives

Volume 8

For further volumes:
http://www.springer.com/series/7906

Landscape Modelling

Geographical Space, Transformation and Future Scenarios

Jiří Anděl · Ivan Bičík · Petr Dostál ·
Zdeněk Lipský and Siamak G. Shahneshin
Editors

Springer

Editors
Dr. Jiří Anděl
Department of Geography
Faculty of Science
Jan Evangelista Purkyně University in Ústí nad Labem
České mládeže 8
40096 Ústí nad Labem
Czech Republic
jiri.andel@ujep.cz

Dr. Ivan Bičík
Department of Social Geography and Regional Development
Faculty of Science
Charles University
Albertov 6
12843 Prague
Czech Republic
bicik@natur.cuni.cz

Dr. Petr Dostál
Department of Social Geography and Regional Development
Faculty of Science
Charles University
Albertov 6
12843 Prague
Czech Republic
dostal@natur.cuni.cz

Dr. Zdeněk Lipský
Department of Physical Geography and Geoecology
Faculty of Science
Charles University
Albertov 6
12843 Prague
Czech Republic
lipsky@natur.cuni.cz

Dr. Siamak G. Shahneshin
SHAGAL/iodaa
Interdisciplinary Office for Design, Architecture & Arts
Zumikerstrasse 3
CH-8700 Kusnacht-Zurich
Switzerland
sgs@shagal.net

ISBN 978-90-481-3051-1 e-ISBN 978-90-481-3052-8
DOI 10.1007/978-90-481-3052-8
Springer Dordrecht Heidelberg London New York

Library of Congress Control Number: 2009942990

Cover illustration: "Approaching the city of Usti via the Elbe River". Photo by P. Raška and M. Balej

Printed on acid-free paper

Springer is part of Springer Science+Business Media (www.springer.com)

Preface

The contemporary community of geographers largely accepts the DPSIR scheme, adopted by the European Environment Agency, which denotes the sequence of variables leading from a factor exerting pressure with a particular consequence in a landscape and its reverse impact feeding back on the initial factor. Such a sequence of causal relationships can be studied at different levels of time and spatial scales. One cycle of the sequence in a specific space results in a differential between two states over a period of time, i.e. a change (Antrop, 2005), and when several such cycles are repeated, a development takes place (cf. Present Changes in European Rural Landscapes by Lipský or Memory of a Landscape - A Constituent of Regional Identity and Planning by Balej et al., this volume) in which there may be turning points that are more or less significant. At the end of the Cold War by the end of the 1980s, a large part of Europe, particularly in the countries in East Central and Eastern Europe, entered a new period of societal transition. This transition included changes in political, social, economic, intellectual and environmental values and it also started to reshape the environment in which the societies concerned are living. At the same time, however, these changes had an impact on other parts of Europe and the whole of Europe as well, as each of its countries had to reflect the new development. The actual changes in the landscape that this process caused at various hierarchical scales form part of the long-term formation processes of the European landscape. With regard to the different time and spatial scales and given the aspects we observe, these changes can be perceived as more or less marked. In any case, the changes document the fact that the landscape is a truly living entity which incorporates countless networks of relations and mechanisms.

In 2004, a team of researchers from the Department of Geography, Faculty of Science, J.E. Purkyne University in Ustí nad Labem, coordinated by Jiří Anděl, made a successful application to start a research project entitled *Methodical Procedures of Social and Ecological Linkages Assessment in Economic Transformation: Theory and Application.* Its purpose was to identify ecological and social aspects of the transition process in the Czech Republic and to propose methodological procedures for its assessment. The processes of landscape changes (ecological and social subsystems of the landscape) and the forces driving these processes, as well as their consequences, were studied in their historical context at several spatial hierarchical levels (country, region, and survey areas in different types of landscape). In 2008, when the project entered its final stage, the team began to prepare an international scientific event to facilitate presentations of different approaches to current landscape research, as well as allow specific discussions concerned with the subject matter in terms of space and time, and intellectual understanding of a landscape as a living entity. A conference entitled *Living Landscape: Memory, Transformation and Future Scenarios* which was held in Ústí nad Labem in November 2008 and attracted a large audience from different parts of the world, for example from the US, South Korea, Switzerland, Austria, Slovakia, etc. This book is a selection of contributions presented at the conference and also includes some other papers relating to the conference issue.

Of course, it is necessary for the publication purpose to give creative and somewhat unrestrained discussions a consistent integrating shape with a comprehensible message. This is why both the title of the book and its parts and contents had to be adjusted. The parts of the book bring together contributions concerned with related subject matter and which are loosely connected with each other. Each part begins with a synopsis posing questions that the papers concerned try to answer. The editors made an effort for each paper to reflect hierarchical levels of the issues being addressed with their specific spatial dimension and a time horizon. The contribution by Siamak G. Shahneshin in the first part, entitled *Where the Moral Appeal Meets the Scientific Approach*, gives an overall framework outlining connections between the transformation of a specific landscape and people's moral bearings, thus unveiling the deeper context of the scientific study of a landscape as a living entity as presented in the subsequent parts of the book. The second part, *The Concept of Landscape in Contemporary Europe,* attempts to look at various ways of interpretation of the landscape as a system, its changes (Zdeněk Lipský) and its possible classification and assessment in contemporary Europe (Jiří Anděl et al.). When Richard Hobbs (1997) speaks of the landscape as the best scale for measuring local effects of global changes, one must add that for an actual landscape and for management and planning policy, it is often essential to conceptually organise landscape components – internally heterogeneous, functionally variable and spatially fluctuating – into regions or localities. Considering the close linkages between natural and social phenomena, impacts in landscape can only be evaluated on a clearly delimited spatial-temporal level, i.e. based upon a conceptual and data framework. Linking landscapes and multi-scale regions is the subject matter of the third part entitled *Between Landscapes and Multi-Scale Regions,* in which the authors are concerned with both regional differentiations in perceptions of selected phenomena at macro scale across the European Union (Petr Dostál) and at regional and local scales of geographic systems considering significance and consequences of their internal transformation (Hartmut Kowalke et al.; Ivan Bičík et al.). Various issues of regions and localities influenced by internal and, particularly, by external forces, are discussed specifically in the fourth part of the book, *The Changing Face of a Landscape: Identity and Perception*, in which the authors are also concerned with reverse effects of specific changes in the landscape and consider the question to what extent a sequence of changes can be understood as a continuum and when and where a turning point begins. The authors look for answers to such questions through analyses of changes in regional identities and social perceptions of the landscape (Martin Balej et al.; Martin Prinz et al.; Milan Jeřábek). Finally, the fifth part, entitled *Modelling and Geovisualisation in Landscape Planning and Management*, is a collection of papers discussing applications of modern technologies to the issues analysed in the preceding parts. The authors deal with the issues of retrospective geovisualisation and future landscape development scenarios for the purpose of landscape planning (Tomáš Oršulák and Pavel Raška), landscape structure analysis for the purpose of sustainable planning (Christa Renetzeder et al.), landscape modelling in biodiversity studies (Stefan Schindler et al.), and geoinformational means of representing selected phenomena in the landscape (Jana Svobodová and Vít Voženílek).

The editors of this book are grateful to all those who participated in its preparation and who made this project happen. At the very beginning, this was the team that cooperated in the above-mentioned research project and organised the November 2008 international conference, supported by a grant from the Ministry of Labour and Social Affairs of the Czech Republic. Acknowledgements are also due to all the participants, of whom some contributed to this book. We thank them all for their efforts and for their consistence in observing the purpose of this book. Its quality was significantly improved by the expert co-editors through comments and recommendations they made. We wish to thank Pavel Raška and Tomáš Oršulák for maintaining communication with the editors of the Springer publishing house and the authors from the very beginning, as well as for the technical processing of the contributions. Last but not least, we would like to thank the Springer team, headed by Geosciences editor Robert Doe, and publishing assistant Nina Bennink, as well as the Series editor, Giovanni Maciocco, and his colleagues and project assistants, Monica Johansson and Lisa Meloni for their tireless help in drafting this book.

Usti nad Labem, Czech Republic — Jiří Anděl
Prague, Czech Republic — Ivan Bičík
Prague, Czech Republic — Petr Dostál
Prague, Czech Republic — Zdeněk Lipský
Zurich, Switzerland — Siamak G. Shahneshin

Contents

Contributors

Jiří Anděl
Department of Geography, Jan Evangelista Purkyne University in Ústí nad Labem
České mládeže 8, 400 96 Ústí nad Labem, Czech Republic
jiri.andel@ujep.cz

Jiří Anděl graduated from Charles University in Prague and specialised in social geography and demography. He was the Head of the Department of Geography, J.E. Purkyne University for 9 years. His research has been mainly on social geography, regional geography and population geography in relation to the environmental aspects.

Martin Balej
Department of Geography, Jan Evangelista Purkyne University in Ústí nad Labem
České mládeže 8, 400 96 Ústí nad Labem, Czech Republic
martin.balej@ujep.cz

Martin Balej obtained his PhD in Faculty of Science of Charles University in Prague. In his research activities he focuses on landscape ecology, landscape assessment methods, landscape metrics, evaluation of land use/land cover change and the use of modern geographical information tools.

Ivan Bičík
Department of Social Geography and Regional Development, Charles University in Prague
Albertov 6, 128 43 Praha 2, Czech Republic
bicik@natur.cuni.cz

Ivan Bičík gained his doctorate at the Charles University in Prague, where he still works now. Former president of the Czech Geographic Society and head of the department, he focuses especially on environmental and regional geography, and land use studies (member of IGU/LUCC Commission).

Alena Chvátalová
Department of Geography, Jan Evangelista Purkyne University in Ústí nad Labem
České mládeže 8, 400 96 Ústí nad Labem, Czech Republic
alena.chvatalova@ujep.cz

Alena Chvátalová obtained her PhD in physical geography from Charles University in Prague. She especially focuses on regional physical geography, landscape potential and risks and geomorphology. She has been the vice-rector of the J.E. Purkyne University in Ústí nad Labem since 2007.

Petr Dostál
Department of Social Geography and Regional Development, Charles University in Prague
Albertov 6, 128 43 Praha 2, Czech Republic
dostal@natur.cuni.cz

Petr Dostál studied geography from 1965 to 1968 at Charles University and settled in the Netherlands in 1968. He graduated in social geography from the State University of Groningen (M.A.), and received his PhD from the University of Amsterdam. He is currently professor at the Charles University in Prague and his research is concerned with regional development, risk processes and European integration.

Milan Jeřábek
Department of Geography, Jan Evangelista Purkyne University in Ústí nad Labem
České mládeže 8, 400 96 Ústí nad Labem, Czech Republic
milan.jerabek@ujep.cz

Milan Jeřábek obtained his PhD in social geography and regional development from Charles University in Prague. Academic career in Faculty of Science of Charles University in Prague, at the Institute of Sociology of Academy of Sciences of the Czech Republic, and he is currently in Faculty of Science of UJEP in Ústí nad Labem, with specialisation in social geography, regional planning and politics, and cross-border issues.

Jan Kabrda
Department of Social Geography and Regional Development, Charles University in Prague
Albertov 6, 128 43 Praha 2, Czech Republic
kabrda@seznam.cz

Being a PhD candidate at Charles University in Prague, Jan Kabrda studies land-use changes in relation to their social and political driving forces as well as regional differences of land-use structure and changes. He focuses his research on Czechia in a Central-European context.

Vassiliki Kati
Department of Environmental and Natural Resources Management,
University of Ioannina
Seferi 2, 30100 Agrinio, Greece
vkati@cc.uoi.gr

Vassiliki Kati is a biologist, who received her PhD degree in biodiversity conservation at the Université Catholique de Louvain (Belgium). Her research focuses on biodiversity assessment and conservation using multi-species data from insects and vertebrates. She is a lecturer at the University of Ioannina (Greece), and board member of the society for Conservation Biology - European section.

Dionisios Kalivas
Laboratory of Soils and Agricultural Chemistry, Agricultural University of Athens
75 Iera Odos, 118 55 Athens, Greece
Kalivas@aua.gr

Dionisios Kalivas is Assistant Professor at the Agricultural University of Athens (Department of Natural Resources Management and Agricultural Engineering). He teaches GIS, Spatial Statistics and Geostatistics. He has been involved in numerous research projects and he is author of more than 50 publications in refereed journals and conference proceedings.

Michiel Kiers
Geo-Information Centre, ALTERRA, Postbus 47, 6700AA Wageningen
The Netherlands
michiel.kiers@wur.nl

Michiel Kiers is researcher at the centre for Geo-Information at Alterra, the Netherlands. His expertise is spatial analysis and modelling in projects oriented to landscape ecology, especially to landscape structure and land cover changes.

Hartmut Kowalke
Lehrstuhl für Wirtschafts- und Sozialgeographie Ost- und Südosteuropas,
Technische Universität Dresden
01062 Dresden, Germany
hartmut.kowalke@mailbox.tu-dresden.de

Hartmut Kowalke has been a member of the Faculty of Forest, Geo and Hydro Sciences since 1992 and Director of the Institute of Geography since 2002. He is a head of Professorship of Economic and Social Geography of East and Southeast Europe. His research activities are focused on regional development of Saxony, East Germany and the European Union and on the trans-border cooperation between Saxony and Czech Republic.

Zdeněk Lipský
Department of Physical Geography and Geoecology, Charles University in Prague
Albertov 6, 128 43 Prague, Czech Republic
lipsky@natur.cuni.cz

Zdeněk Lipský is a landscape ecologist and geoecologist who received his doctorate at the Charles University in Prague. In his research he deals with landscape change, typology and assessment in relation to the overall face of a landscape as well as to its individual functions.

Katja Lohse
Lehrstuhl für Wirtschafts- und Sozialgeographie Ost- und Südosteuropas, Technische Universität Dresden
01062 Dresden, Germany
katja.lohse@mailbox.tu-dresden.de

Katja Lohse has been a member of the Technical University of Dresden, Faculty of Forest, Geo and Hydro Sciences since 2008. She works at the Department of Economic and Social Geography of Eastern and South-eastern Europe. Her research interests are focused on the development of city structures in European, former socialistic states as well as the cross-border cooperation in Euroregion Elbe/Labe.

Sander Mücher
Centrum voor Geo-Informatie (Centre for Geo-Information), ALTERRA
Droevendaalsesteeg 3, 6708 PB Wageningen, The Netherlands
sander.mucher@wur.nl

Sander Mücher is a researcher at the centre for Geo-Information at Alterra, the Netherlands. He focuses on the development of new techniques and methods in the field of habitat, land cover and landscape monitoring and the integration of remote sensing with additional geographic information and models.

Jiří Najman
Department of Social Geography and Regional Development, Charles University in Prague
Albertov 6, 128 43 Praha 2, Czech Republic
najman@natur.cuni.cz

Jiří Najman is a PhD candidate at Charles University in Prague. In his research he deals with land-use changes and application of GIS methods and use of remote sensed images in landscape studies. Terriotorially, his research is primarily aimed at Central Europe and the area of former Iron Curtain.

Tomáš Oršulák
Department of Geography, Jan Evangelista Purkyne University in Ústí nad Labem
České mládeže 8, 400 96 Ústí nad Labem, Czech Republic
tomas.orsulak@ujep.cz

Tomáš Oršulák is a lecturer at the Department of Geography in the Faculty of Science UJEP (since 2001). Presently, he is a PhD candidate at the Institute of Geoinformatics, Technical University Ostrava. He specialises in geographic informational systems, geovisualization, 3D modelling and application of virtual reality (CAVE system) in landscape and territorial planning.

Aristotelis Papageorgiou
Department of Forestry, Environment and Natural Resources, Democritus University of Thrace
Pantazidou 193, 68200 Orestiada, Greece
apapage@fmenr.duth.gr

Aristotelis Papageorgiou received his PhD degree in forest genetics at the University of Göttingen (Germany). He is chair of the Forest Genetics Laboratory at the Democritus University of Thrace (Greece). He also developed activities in forest and environmental policy and he acted as an EU and national delegate in the UN and the FAO.

Kostas Poirazidis
WWF Greece Dadia project
68400 Soufli, Greece
ecodadia@otenet.gr

Kostas Poirazidis studied forestry and environmental protection in Thessaloniki. He received his PhD degree in raptor habitat modelling and conservation. His main interests are biodiversity conservation, management of natural resources and ecological modelling. Since 2003, he teaches at the Democritus University of Thrace and at the Technological Education Institute of the Ionian Islands.

Martin A. Prinz
Department of Conservation Biology, Vegetation & Landscape Ecology, University of Vienna
Rennweg 14, A-1030 Vienna, Austria
martin.prinz@univie.ac.at

Martin A. Prinz is a graduate Ecologist and PhD candidate at the University of Vienna. Since the beginning of 2005, he has been working on several national projects dealing with landscape structure, indicators for sustainable landscape development and tools for the assessment of environmental effects of land use and agri-environmental subsidies.

Pavel Raška
Department of Geography, Jan Evangelista Purkyne University in Ústí nad Labem
České mládeže 8, 400 96 Ústí nad Labem, Czech Republic
pavel.raska@ujep.cz

Pavel Raška is lecturer at the Department of Geography in the Faculty of Science UJEP. Precently, he is a PhD candidate in the Geographical Institute, Masaryk University in Brno. In his research he focuses on palaeogeomorphology and environmental change of rock-mantled slopes, biogeomorphic systems in a landscape, geomorphic risks, historical geomorphology and long-term landscape changes.

Karl Reiter
Department of Conservation Biology, Vegetation & Landscape Ecology, University of Vienna
Rennweg 14, A-1030 Vienna, Austria
karl.reiter@univie.ac.at

Karl Reiter is Assistant Professor at the University of Vienna. During the last years he tried to develop strategies in sampling design based on spatial factors manly derived from Digital Elevation Models and classification of remote sensed data.

Christa Renetzeder
Department of Conservation Biology, Vegetation & Landscape Ecology, University of Vienna
Rennweg 14, A-1030 Vienna, Austria
christa.renetzeder@univie.ac.at

Christa Renetzeder is a PhD candidate at the University of Vienna. Since 2005, she has been working with landscape structure, indicators for sustainable landscape development and tools for the assessment of environmental effects of land use.

Stefan Schindler
Department of Conservation Biology, Vegetation & Landscape Ecology, University of Vienna
Rennweg 14, A-1030 Vienna, Austria
stefan.schindler@univie.ac.at

Stefan Schindler is a research assistant at the Department of Conservation Biology, Vegetation Ecology and Landscape Ecology (University of Vienna). He is currently finishing his PhD on landscape and biodiversity pattern. His main research foci are landscape ecology, biodiversity research, agricultural policy, and sustainable forest management.

Olaf Schmidt
Lehrstuhl für Raumordnung Technische Universität Dresden
01062 Dresden, Germany
olaf.schmidt@mailbox.tu-dresden.de

Olaf Schmidt has been a member of the Faculty of Forest, Geo and Hydro Sciences since 1992. He works at the Institute for Geography. His special subjects are spatial and regional planning. The research activities are focused on regional development of Saxony and on the trans-border cooperation between Saxony and Czech Republic.

Siamak G. Shahneshin
SHAGAL | iodaa, Interdisciplinary Office for Design, Architecture & Arts
Zumikerstrasse 3, CH-8700 Küsnacht-Zurich, Switzerland
sgs@shagal.net

Siamak G. Shahneshin is Professor of urban planning, ecological landscape architecture, and sustainable architecture. Trained at the Accademia di Belle Arti Firenze, and Politecnico di Torino, GSD Harvard, Architectural Association London, ETH Zurich. Prof. Shahneshin worked with many renowned architects before he co-founded SHAGAL | iodaa, based in Zurich, concerned with issues of urban growth, presenting new problems related to land use, spatial and economic organisation.

Jana Svobodová
Department of Geoinformatics, Palacky University in Olomouc
tr. Svobody 26, 771 46 Olomouc, Czech Republic
svobodova-j@centrum.cz

Jana Svobodová works as a lecturer in Geoinformatics at Palacky University Olomouc in the Czech Republic. She specializes in digital elevation models and application of Geographical Informational Systems in geomorphology. Her recent interests is related to analyses of precision of digital elevation models.

Michiel van Eupen
Centrum Landschap (Landscape Centre), ALTERRA
Postbus 47, 6700AA Wageningen, The Netherlands
michiel.vaneupen@wur.nl

Michiel van Eupen is researcher at the landscape centre at Alterra, the Netherlands. He has extensive experience with spatial analysis and implementation of landscape ecological concepts into models and landscape indicators for risk and sustainability assessment.

Henrik Von Wehrden
Institute of Biology - Geobotany and Botanical Garden, Martin-Luther-University Halle-Wittenberg
06108 Halle, Germany
henrikvonwehrden@web.de

Henrik Von Wehrden is a trained geographer with a strong background in vegetation science. He aims to combine spatial information (including ground truth data, remote-sensing products, modelled layers etc.) and statistical analyses to derive key data and results for nature conservation.

Vít Voženílek
Department of Geoinformatics, Palacky University in Olomouc
tr. Svobody 26, 771 46 Olomouc
Czech Republic
vit.vozenilek@upol.cz

Vít Voženílek is a Professor in Geoinformatics at Palacky University Olomouc in the Czech Republic. His research relates primarily to modelling in GIS and thematic and atlas digital cartography. He is a member of IGU Commission on GIS and ICA Commission on National and Regional Atlases.

Thomas Wrbka
Department of Conservation Biology, Vegetation & Landscape Ecology, University of Vienna
Rennweg 14, A-1030 Vienna, Austria
thomas.wrbka@univie.ac.at

Thomas Wrbka is Assistant Professor at the University of Vienna with expertise on landscape classification, concepts for sustainable land use, analysis of correlation between land management and biodiversity, vegetation and landscape monitoring as well as the development of management concepts.

Part I
Where the Moral Appeal Meets the Scientific Approach

What makes our world exist in a state of crisis? How can the expansionist's thinking be changed? What does the shrinkage concept refer to? How can one apply it in landscape planning? How is the system design approach applied in landscape transformation towards sustainability?

Chapter 1
The Weeping Landscape

Siamak G. Shahneshin

1.1 Rising Bubble

With this contribution, I would like to raise an urgent question: How can our world best avoid committing ecological suicide?

Whether you accept it or not, we have crafted a culture bubble, and built an environ bubble, within which the mindset of expansionistic thought represents the overculture. The challenge today is to deflate the bubble before it bursts. The most vulnerable sector may be the environ in the extended sense of the word (cf. Rees, 2003; Diamond, 2005). Whether you agree or disagree a bee without honey is a simple illustrative example of the very nature of today's design culture.

In each epoch, expansionistic thinking has been both creative and destructive, but today it is the very existence of humanity, and the planet, which is at stake. Expansionism is all about satisfying individual wants, while society requires sublimating one's desires (and the willingness to compromise).

Conversely, the basic point of *shrinkage* (Shahneshin, 1996, 2004, 2008d) is that sooner or later our principle premises concerning growth and expansion must be urgently revised and reassessed. *Shrinkage* is global in reach, ranging from the well-being of nature (Shahneshin, 2008a) to finance, from families to cities, and so on, yet *shrinkage* is still in an embryonic stage. Needless to say, time is running out. We need to act at wartime speed (Shahneshin, 2007a).

As a result, one of the best places to seek understanding of *shrinkage* is in the study of sprawl (Hirschhorn, 2005) and postsprawl and the devastating implementation of those modern, and post-modern theories, as well as present hyper-thinking trends which share their eudaemonist concerns. Given the systematically disappointing results of these approaches, it is time to look seriously at the alternatives. Ecological Landscape Urbanism (Shahneshin, 1996, 1998) is a catalyst leading towards a sustainable world (Shahneshin, 2004, 2006a, 2007b, 2008e).

S.G. Shahneshin (✉)
SHAGAL | iodaa[1], Interdisciplinary Office for Design, Architecture & Arts, Zumikerstrasse 3, CH-8700 Küsnacht-Zurich, Switzerland
e-mail: sgs@shagal.net

J. Anděl et al. (eds.), *Landscape Modelling*, Urban and Landscape Perspectives 8,
DOI 10.1007/978-90-481-3052-8_1,

1.2 Climate Change and Landscape

As I leave the mountains of Engadin on this warm Autumn morning, my mood swings between hope and gloom. I'm happy to have witnessed that over the last six months of 2008, environmental landscape planning awareness has seen a wealth of global seminars and conferences showing that every country in the world is willing to make changes that will have a positive effect on the world. These months saw "environmental planning and landscape ecology" often discussed in magazines and a flaunting with new European regulations. But at the same time the figures I had read throughout the last six months of 2008 disturb me greatly (Shahneshin, 2008b).

In nature, one-way linear flows do not survive long. Nor, by extension, can they survive long in the expanding economy that is not a part of the earth's ecosystem. The challenge is to redesign economy and development so that they are compatible with nature. The throwaway economy and runaway development that have been evolving over the last half-century are an aberration, as can be seen by the collapse of financial systems worldwide in October of 2008 (Shahneshin, 2008c).

There is no doubt that, as our built environment has transformed from a local phenomenon to a global one, we are now confronted with more pressing social, technological, economic, environmental and political change forcing us into a local mindset – on a global scale (Shahneshin, 2008d; Stern, 2006).

We are living in an epoch capable of building the most extraordinary infrastructures, but these same projects have seldom been able to structure the territory that they traverse and occupy. Since SHAGAL | iodaa[1] is in the business of design, it has made great efforts to address this very issue in its extended sense; leading city administrators and policy-makers in creating a city where the built and natural environments prosper and thrive "together" (Shahneshin, 1996).

SHAGAL | iodaa has, since the early 1990s, embarked upon hypothetical enactments of a city carbon-neutral policy for numerous projects including the Cincinnati Park in Torino (Italy) 1994, Strategic Masterplan for Downtown Athens (Greece) 1998, Masterplan for a New City in the Eastern region of China 2002, Trinity River Corridor Development in Dallas (USA) 2003, Riverfront Development in Geneva (Switzerland) 2004, the New Masterplan for Zurich Airport (Switzerland) 2005, and for the Hobart Waterfront in Tasmania (Australia) 2006, to name some.

1.3 Shrinking Airport

I would love to share with you one of the mentioned projects. The greatly discussed Zurich Airport New Master Plan project: a truly participatory approach of nature and men. Before telling you the story of this master plan (the Naturpark), it would be compelling to reveal the bottom line and foremost imperative engines of this neighbourhood- and community-oriented project.

People and nature are placed at the heart of this design with quality *shrinkage*[2] as the main programmatic theme, and it is called the *new* "smart growth", adding to the discourse surrounding urban landscape in Europe and beyond.

Further consideration reveals the impossibility of adequately conceiving the airport as either a building or an urban ensemble. What is an airport if not a contiguous, highly choreographed, scrupulously maintained and regularly manicured landscape? In revisiting the site of the contemporary airport, SHAGAL | iodaa's work examines one of the most emblematic sites of contemporary urban[isation], re-framing it as an enormous public landscape.

This re-framing of the landscape offers extensive value to the discipline of landscape architecture and land planning, creating a critical space for the examination of the contemporary city and the role of the designer/decision-maker within it. In so doing, this work offers a cultural framework for intervention in sites of contemporary urbanisation. For many, *shrinkage* alone seems capable of rendering the contemporary city's order, scale, and lack of density, both social and spatial. By focusing design intelligence and research attention on the status of landscape in the contemporary city, this work recommends itself for further reading by audiences local and remote.

Contemporary landscapes are challenged by economic realities of a new kind, which create mutant environments that transform sites and adapt them to the whims and exigencies of complex infrastructures and logistics. The environmental complexity of such sites is overwhelming, in terms of visual aesthetics first, but also in terms of cultural and environmental understanding and integration.

This particular landscape intelligence is new, because there are no past references for such environments. Zurich airport was not conceived as a landscape per se, but rather as a large piece of infrastructure permitting machines to land and take-off.

The review of Zurich airport and recent economic and social events led to critical attention being paid to *shrinkage*. Reinstating and maintaining the flora and fauna in this area – instead of expanding the airport – required a "whole systems" design approach. Zurich airport is a territory in itself, an island with all its rules and regulations. The "choreographic" dimension not only has a direct impact on the site, but also across the entire region. The airport generates both value and disvalue. We have reached a paradox in landscape – and land planning – which we are no longer able to operate upon.

The [re]invention of nature along those narrow lines becomes a challenge for a whole generation of landscape architects to come. SHAGAL | iodaa, unlike many, didn't tackle land (or landscape) at a scale that has remained until now very abstract and distant. Talking in Coleridge language, we have to say that SHAGAL | iodaa's design creates an endless text, an endless translation of the original that is aware of its contradictoriness.[3] The aim is to be as true to the original as possible, that is, to make viewers forget that the landscape *tableau* is really not as rigidly eternal as the painting stored in the cultural memory.

The former site of Zurich airport was entirely woodland and hosted a diverse array of rare vegetation, so-called "Swiss Natural Good". It was the home habitat to 316 species that thrived in these landscapes before men, in the mid-1960s, bulldozed it into an alien district like an omelette scrambled out of existence causing widespread changes in vegetation patterns, distortion of the Glatt river and disconnection of natural reservoir areas. A consequence of this was that the number of species has been greatly reduced and currently there are only 22 species living there.

The design for the ambitious endeavour to transform Zurich Airport's contaminated land into parkland was not easy at all from the beginning. SHAGAL | iodaa offer a longer term strategy based on natural processes and plant life cycles (successional development) to rehabilitate the severely degraded landscape. Surprisingly, these areas provide a regionally significant wildlife sanctuary for diverse species of animals. SHAGAL | iodaa envision a rich reservoir not only for wildlife, but also for cultural and social life, restoring existing grasslands, patches, forests, and reinvigorating the rare species of vegetation while introducing new habitats and adding amenities for learning from flora and fauna.

The entire new master plan (the Naturpark), from the beginning (mid 2002) up to the final presentation (late 2005) is based on facts: Zurich airport's financial failures, functional and technical fiascos as well as the high number of accidents per year.

SHAGAL | iodaa's members have interviewed over 250 people, one-on-one, who live and work in the vicinity of Zurich airport, including citizens and authorities of the eight neighbouring cities. This was accomplished through house-to-house visits and questionnaires, collecting data, demonstrators' resolutions,[4] historical plans, flora and fauna along with statistics etc., organising community charrettes (workshop conversations) and symposium-type forums.

Planning by listening to the landscape and its users – the core of SHAGAL | iodaa's thinking – is so logical that it's almost impossible to plan differently. So, despite the fact that the airport management had planned to expand the airport and the expansion plans were ready, SHAGAL | iodaa embarked on a redesign of the entire airport and its neighbourhood (without a commission, SHAGAL | iodaa's founders felt the need to reconsider the plans and acted accordingly). We embarked upon a design programme. It is not only a physical programme; it is also a political, economic and environmental programme that allows things to happen, a bottom-up form of Ecological Landscape Urbanism that distances itself from authorship or trademark control over form, while allowing for specificity and responsiveness to the environment.

SHAGAL | iodaa designed a *shrinkage* for the airport reversing the usual approach to airport design, a re-naturalisation of the territory placing priority on open space and natural systems rather than on buildings and infrastructure (Figs. 1.1 and 1.2). This master plan proved that the airport can function efficiently at a high capacity within a smaller boundary. The FOCA (Swiss Federal Office of Civil Aviation) rightly bans expansion plans for at least 25 years, in order to avoid further accidents in this area.

The new master plan for Zurich Airport, is a multi-staged approach that evolves over time, allowing a slow [re]generation of the degraded place into a quintessential eco-aesthetic landscape, with a dynamic staging offering both indeterminacy and uncontrolled occupation in four major design phases – which seeks to evolve over time. The Ecological Landscape Urbanism approach under the *shrinkage* umbrella is therefore not only concerned with being ecologically correct, but also anthropologically correct in a place where nature has become drastically impoverished amidst a weakened urban environment and learning how to work with it creatively. As a

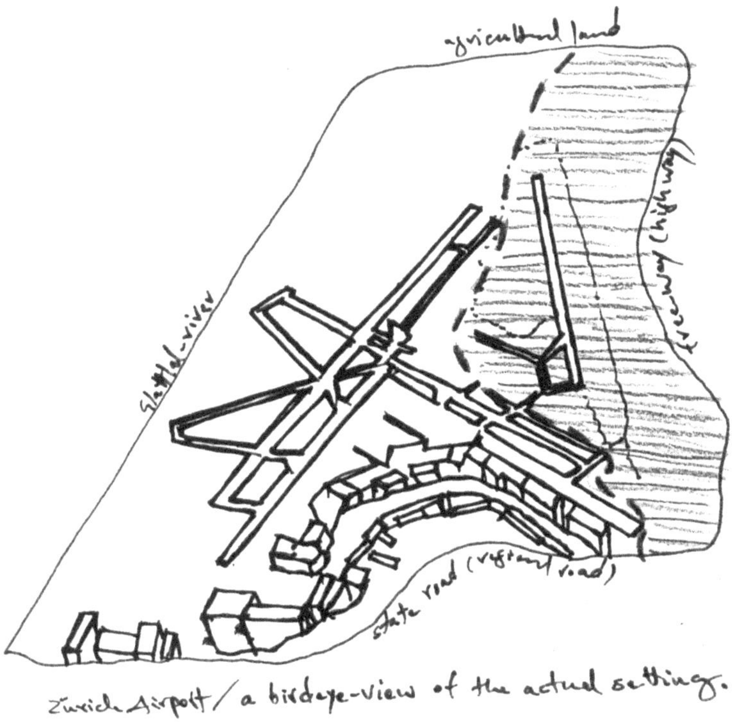

Fig. 1.1 A birds-eye view of the current setting. The *hatched area* refers to the 1st phase of the first *shrinkage* stage

process, the Naturpark in Zurich represents an ecological strategy of environmental reclamation at both natural and social level.

There are four general phases to complete the whole master plan. In the first phase, SHAGAL | iodaa plan to shrink the east runway, creating a Naturpark which opens a natural reservoir to the public as a learning venue of flora and fauna. Unlike green spaces of earlier generations, today's facilities should not be passive land intended for communing with nature only. This Naturpark seeks to engage people, intellectually and physically. Additionally, this design concept could [re]store the Kloten areas' biodiversity, replenish ecological habitats, boost tourism and job creation and protect drinking water supply catchments instead of polluting drinking water and sending many more species into extinction and negatively affecting tourism. SHAGAL | iodaa's design for the Zurich airport area shows how one region could reconsider the value of its natural capital to benefit both the local economy and the global community by adapting *shrinkage* values.

Its most powerful contribution, however, may be that it recalls nature's restorative cycles and puts them back to work in the city and beyond. The real winner of this *shrinkage* proposal would be the environment: a treasure trove of natural wealth will be accessible as a pedagogical medium – in changing user behaviour through

Fig. 1.2 *Shrinkage* stage one, New Master Plan of Zurich Airport[5]

Fig. 1.3 Elevated wooden paths allow users to experience the landscape

education and awareness, and to support also the region's characteristic biodiversity. The income would be clearly visible to Zurich's neighbourhood residents and visitors, demonstrating the positive local (and global) contribution this Naturpark would make to world climate. In short, this design is environmentally restorative, socially constructive and economically viable.

Figures 1.2 and 1.3 capture the character and spirit of the new park. The park will be phased-in in four stages over 60 years as sections of the environment must be rehabilitated. Also the whole project will go through regional public referendum.

I foresee the current practice of airport design being abandoned soon, because we will no longer need kilometres of runway area, as we now have manufactured prototypes of civic airplanes that take-off (and land) vertically. Consequently, SHAGAL | iodaa envision that the whole Zurich airport will shrink around 2080.

Notes

1. SHAGAL | iodaa is the official name of the International multidisciplinary collaborative studio for place-responsive programming, research, criticism, writing, teaching and designing (under the shrinkage umbrella) while fuses architecture, landscape architecture, urbanism and the visual arts, founded by Siamak G. Shahneshin and Lui Galati.
2. The terminology *shrinkage* was coined for the design and planning disciplines by Siamak G. Shahneshin in the early 1990s. *Shrinkage* has been proposed to denominate a widespread response to planning, in the extended sense of the word. *Shrinkage* is a way of thinking, and signifies the possibility that humans and other forms of life will flourish on the earth forever. The *shrinkage* concept is pleasingly simple; it's a call to turn the traditional practice of architecture and planning, policy-making and programming (in an extended sense of the word *planning*, for instance, environmental design) inside out placing priority on natural systems. Perhaps we should not think of *shrinkage* as being opposed to growth, rather we can view *shrinkage* as being a facilitator of growth, a sustainable growth. Why consider Zurich airport *shrinkage*? Those who are concerned about it often cite alarming figures. For example, we are told that the USA is losing nearly 400 thousand m^2 of open space to new development each hour, and that Switzerland is losing farmland and forest at the rate of 400 m^2 per hour. Those numbers are so terrifying that it is little wonder that loss of open space has become a top issue among many citizens.
3. "Contradictoriness", refers to the contradiction that Zurich airport's machinery is located in a place that used to be a glacial basin and that this was followed by the intervention of man and the spending of 700 million Euros of public money to replace lost rare vegetation. Several species have become extinct by being moved from their original location.
4. "Demonstrator's resolutions", refers to the resolutions or written requests by people who live near to Zurich airport. These people – the Glattal-Stadt citizens – have organised many demonstrations and several associations have been set up to fight Zurich airport's plans for expansion and the problems caused by Zurich airport.
5. The existing runway 14/32 becomes part of a united natural reservoir. Temporal urbanism with different and various uses such as installations, public art, markets and events characterise the old runway and mark the backbone of the site. A series of linear elevated paths with low maintenance make the previously "forbidden" natural reservoir area accessible. They create a pattern of fields with a variety of plantations and minimum maintenance strategies. These paths sometimes intersect and cross the existing highlighted ground paths. And the path system along the runway is made accessible through these new elevated wooden paths.

References

Diamond, J. M. (2005). *Collapse: How societies choose to fail or succeed.* New York: Viking.
Hirschhorn, J. S. (2005). *Sprawl Kills.* New York: Sterling & Ross Publishers.
Rees, M. J. (2003). *Unsere letzte Stunde.* München: Bertelsmann Verlag.
Shahneshin, S. G. (1996). L'irrazionalità del razionale. *Bioarchitettura, 15*(6), 4–5.
Shahneshin, S. G. (1998). La ricerca dell'ecologia perduta. *Bioarchitettura, 17*(10), 5.
Shahneshin, S. G. (2004). *Shrinking smart.* Lecture held at Die Eidgenössische Technische Hochschule, Zürich.
Shahneshin, S. G. (2006a). *Planners, listen to the City!* Lecture held at University of New Mexico, Albuquerque, NM, USA.
Shahneshin, S. G. (2006b). Walk the talk. In: *Sustainable development.* Hong Kong: Hong Kong University Press.
Shahneshin, S. G. (2007a). *Lilliput or brobdingnag.* Lecture held at University of Portsmouth, Portsmouth.
Shahneshin, S. G. (2007b). Lege das lexikon beiseit. In: *Alles wird gut.* Lüneburg: Universität Lüneburg Verlag.
Shahneshin, S. G. (2008a). La natura, la nostra guida. In: G. Marucci (Ed.), *Architettura oltre la forma.* Milano: Di Baio Editore.
Shahneshin, S. G. (2008b). It will affect life on earth. *Landscape, 18*, 26–28.
Shahneshin, S. G. (2008c). Knowing nature. *Landscape, 16*, 18–22.
Shahneshin, S. G. (2008d). A manifesto for better world. *Landscape, 15*, 20–23.
Shahneshin, S. G. (2008e). Learning from flora & fauna. *Landscape, 13*, 44–46.
Stern, N. H. (2006). *Economics of climate change*, London: British Royal HM Treasury Ministry.

Part II
Landscape Concept in Contemporary Europe

How is the term landscape understood in Europe? What are the basic mechanisms of landscape changes? How does the new wilderness evolve in contemporary Europe? How do ecological and social factors interact in landscape development? What is the environmental stress? What are the integrative methods for landscape assessment?

Chapter 2
Present Changes in European Rural Landscapes

Zdeněk Lipský

2.1 Topical Issue of Landscape Changes

Landscape changes represent an extremely wide as well as very important and topical issue in landscape sciences. The number of papers in scientific journals that focus on the topic of landscape changes has been increasing explosively during the last two decades (Aspinall, 2006). Among many conferences, workshops and seminars dealing with the topic of landscape changes, the seminar *Landscape change and its ecological consequences in Europe* held in Tilburg in 1995, from which the important report on the state of land use and landscape change in Europe in the 1990s was published (Jongman, 1996), should be mentioned. The importance of recent landscape changes and their consequences are further discussed in the monograph edited by Mander and Jongman (2000). The international seminar organised in Norwegian Tromso in June 2006 has a concise title: *Landscape Change: Learning from the past – Visions for the future.*

Landscape is a theme in many disciplines, resulting in diverse approaches (Antrop, 2008). The fast changes occurring today have caused the growing popularity of landscape itself and landscape changes in particular. A growing public and political interest in landscape issues has resulted in the adoption of the European Landscape Convention (Council of Europe, 2000). The great merit of the Convention is that it initiated many programmes for studying landscapes in most European countries as well as on the Pan-European level as never before (Antrop, 2008). The requirement to identify landscape types, to analyse their characteristics and the forces and pressures which affect them as well as to take note of changes in European landscapes is stressed in Article 6 (Specific measures) of the Convention.

Land use as well as general landscape changes are studied in the fields of both geography and landscape ecology, apart from other scientific and applied disciplines dealing with landscape issues. In the framework of the International Geographical

Z. Lipský (✉)
Department of Physical Geography and Geoecology, Charles University in Prague,
Albertov 6, 128 43 Prague, Czech Republic
e-mail: lipsky@natur.cuni.cz

J. Anděl et al. (eds.), *Landscape Modelling*, Urban and Landscape Perspectives 8,
DOI 10.1007/978-90-481-3052-8_2, © Springer Science+Business Media B.V. 2010

Union, the LUCC (Land Use/Cover Change) Working Group is actively working to follow up land-use changes around the world (Himyiama, Mather, Bičík & Milanova, 2005). Historical land use and landscape-structure changes are studied using old cadastral and military maps, aerial photographs, statistical data on historical land use and other sources of data (Lipský, 2000). Research of historical land use has been widely developed in the Czech Republic (Bičík, 1998; Bičík & Jeleček, 2003; Kolejka, 2002 and many others) as well as in other countries of Central Europe (Gabrovec & Petek, 2003; Krausmann, 2001; Olah & Žigrai 2004).

Land use and landscape structure changes are directly linked to changes in landscape character. In recent years, landscape character assessment (LCA) has become a topical issue of applied landscape science. It is recognised as an important tool for policy-makers and stakeholders to reach a sustainable management of land. In the Czech Republic, the term *landscape character* was officially introduced in 1992 in the Legal Act No. 114/1992 Sb., on nature and landscape protection. Since that time six scientific conferences dealing with landscape character assessment and protection (the last one in February 2009) have been organised and intensive discussions among the scientific community have been running in the country. Several methodological guidelines on LCA have been elaborated and LCA has become a legal instrument of the nature and landscape protection of the state in the Czech Republic. The international project ELCAI (European Landscape Character Assessment Initiative) reviewed state-of-the-art LCA at the national and European level (Wascher, 2005).

2.2 Importance of Land Use and Landscape-Structure Changes from the Point of View of Landscape Ecology

Landscape ecology in its dynamic concept is dealing with three main subjects in the landscape: (1) structure; (2) functions and processes; (3) changes and developments. These main general attributes of every landscape are mutually connected by a complex system of feedbacks (Fig. 2.1).

One of the most important notions is that the landscape structure strongly influences ecological processes and characteristics. Functions and all processes running in the landscape depend directly on and arise from landscape structure, this means from the spatial composition of landscape segments. The pattern is an important feature if one studies the relationship between the various horizontally arranged complexes of landscape elements (Zonneveld, 1995).

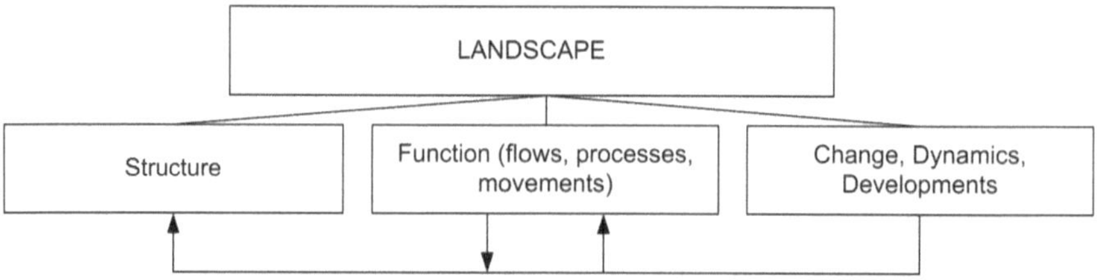

Fig. 2.1 Three main subjects of interest in landscape science in the landscape

Forman and Godron (1986) formulated seven main principles of landscape ecology concerned with landscape structure, landscape functions and landscape change. All the principles lay stress on the primary and absolutely determinant role of landscape structure. According to these main principles, land use and landscape-structure changes have a decisive influence on:

- flows of matter and energy in the landscape;
- flows (movement) of species and information;
- biodiversity and ecological stability of the landscape;
- landscape character, aesthetics and perception of the landscape.

Any changes in landscape structure result in a modified functioning and changed characteristics of the landscape. That is why the study of landscape structure, its changes and consequences represents a crucial issue in landscape ecology.

The main concepts of landscape structure cover the "geocomplex" model and the "patch-corridor-matrix" model as well as the main spatial processes involved in the process of land transformation conceived as changes in the arrangement and spatial composition of the so-called land mosaic (Pietrzak, 2001). Horizontal landscape structure is studied and mapped on different space hierarchical levels from local to regional and global ones depending on the scale and the purpose of the research. We can investigate on the one hand landscape "macrostructure" based on statistical data on land use and land cover and landscape microstructure based on methods of field mapping or interpretation of aerial photos and satellite images on the other hand (Lipský, 2000). The concept of landscape "microstructure" is concisely aimed at the space composition of landscape segments, their mutual relations and connections as well as individual parameters of single landscape components.

Another approach used in landscape typology and landscape character assessment consists of a differentiation between primary, secondary and tertiary landscape structure. The primary structure is determined by natural conditions, i.e. by geological grounds and soils, geomorphological forms, climatic conditions, waters and natural vegetation. The secondary landscape structure can be identified with land use or land cover of the contemporary landscape. Both primary (natural) and secondary (anthropogenic) landscape structures have a direct reflection in the face of the landscape. As the tertiary landscape structure we understand spiritual, immaterial characteristics of the landscape like landscape history and memory, traditions, cultural and historical events as well as various legal restrictions and limits which contribute to the specific landscape character but have got no direct physiognomic expression in the landscape (Lipský, 2008; Mücher et al., 2003).

2.3 Character of Changes in Cultural Landscapes

Landscapes are very dynamic in structure, functions and spatial pattern. Change is inherent to the concept of cultural landscape which is a meeting ground between past, present and future as well as between natural and cultural influences. Landscape dynamics are the basis of landscape diversity and identity (Antrop,

2008). Cultural landscape has been many times likened to the mirror reflecting the state and changes in the society. Changes in society, whether of social, economic, demographic or political character as well as technological progress are more or less reflected in the face of the cultural landscape (Lipský, 1995). Characteristic is the increasing speed and magnitude of the changes. It is a result of the dominant role of man in cultural landscapes.

Landscapes and landscape structures are changing all the time. It concerns both natural and cultural landscapes; change is an intrinsic feature of each landscape. Landscapes have always been adapted to changing needs and technologies (Mander & Jongman, 2000). Björklund (1996) discusses how to interpret landscape as a continuous process of flows and interactions between natural and human-induced processes. The flows are forming and permanently changing landscape structure(s). Landscape changes are running on very different time scales which range from seconds and minutes to long-term changes lasting hundreds, thousands and even more years (see Table 2.1).

Disturbances and changes in landscapes are an intrinsic factor of their existence and development. Since most landscapes are a by-product of human activities they are particularly vulnerable to changes. This is an important characteristic of cultural landscapes that should not be viewed negatively (Meeus, 1995). In cultural landscapes the disturbance regime is dominated by changing land-use practices. Agricultural as well as other cultural landscape types are among those that change most rapidly. Man is the main driver of changes and developments in cultural landscapes. He decides on the method of landscape use, spatial arrangement of ecosystems and their changes. It is significant that anthropogenic processes are

Table 2.1 Time dimensions of landscape-forming processes

Time dimension	Processes
10^6 years	Geological platform tectonics; biological species evolution
10^5 years	Macroclimatic processes (glacials, pluvials); development of relief macroforms
10^4 years	Macroclimatic processes, macrogeomorphology (secular erosion)
10^3 years	Soil formation and development (podsolisation, lateritisation); geo-hydrological processes, long-term successions
10^2–10^1 years	Processes of sedimentation (coastal, fluvial); biological feedback – succession after catastrophes and disturbances; biological invasions; *forestry*
10^{-1}–1 years	*Agriculture, horticulture, urbanisation*
Months	Biological epidemics (diseases), seasonal climatic and vegetation changes, species migrations, *gardening, construction*
Days to hours	Catastrophes caused by meteorological extremes (floods, typhoons, gales, . . .), volcanic activity (eruptions); landslides; *accelerated soil erosion and sedimentation*
Minutes to seconds	Earthquake; avalanches; rock caving, *nuclear explosion*

Anthropogenic processes are distinguished by italics.
Source: Zonneveld (1995) and Lipský (2000)

(on average) much faster in comparison with the course and speed of the majority of natural processes. Fast changes in land use and landscape structure are a distinctive attribute of contemporary cultural landscapes under the dominant influence of man.

Any change in society, whether economic, in ownership, technological or demographic, produces changes in the method of landscape use, in landscape structure and as a result changes in landscape character, biodiversity, ecological stability and in the course of all processes running in the landscape (see above). As societal changes are with time becoming faster, also landscape changes have a tendency to be faster and deeper with more significant ecological consequences. The increasing speed and extent of the changes belies time dimensions of natural development and adaptability of natural systems. Important is here the link made between the transformation of the landscape and the loss of richness and diversity which are considered as characteristic for the European continent and identity (Antrop, 2008).

Brassley (1997) proposed the concept of the ephemeral landscape. Within the relatively stable structure of the landscape, the ephemeral landscape is more or less permanently changing. It is undisputable that changes in agricultural technologies produce changes in agricultural landscapes. Human-induced ephemera are usually associated with agriculture, principally because agriculture is the major form of land use in Europe. The method of cultivation, structure of field crops, harvesting methods, whether of grass or corn, methods of livestock farming as well as other agricultural processes have been radically altered during the last 50 years with concomitant effects on the ephemeral landscape structure. The appearance of the countryside during the corn or hay harvest has been fundamentally changed. Black-and-white photographs from the mid-Twentieth century show ephemeral elements typical of the rural landscape of past centuries that no longer exist in the contemporary landscape. Instead of the lines of shocks that covered the cornfields often for several weeks in the summer season, bales of straw of different size and shape (depending on used technologies) are typical for the present agricultural landscape in late summer. Thus, we can find numerous landscape features that are ephemeral, some natural, some produced by human activities. Brassley (1997) argues that ephemeral components and ephemeral changes have a major impact on the appearance of the landscape and on the way in which it is perceived and valued.

2.4 Socialist Collectivisation as an Example of Dramatic Landscape Changes

The socialist collectivisation of agriculture that has been occurring since the 1950s in Central and Eastern European countries of the former Soviet block has been often presented as a typical example of fast and dramatic landscape-structure changes caused by major political, social and economic changes in the life of a society. There have been many land use and landscape-structure changes throughout history, but those that have occurred since the 1950s have no equivalent in terms of their speed and extent in the Czech rural landscape. According to official instructions, parcels of arable land were unified so as not to be interrupted by meadows, pastures, shrubs or other elements hampering efficient cultivation. During the transition to socialist large-scale production, landscape structure changed rapidly towards its significant simplification (Lipský, 1991). The traditional fine-grained structure of the Czech rural landscape corresponding with small-scale private agriculture technologies changed dramatically and non-reversibly during that time. The size of agricultural holdings was increased 50-times, many meadows in floodplains were ploughed and most of the permanent vegetation structures in the open agricultural landscape were removed (Lipský, 1995). Agricultural plots were perceived as only a monofunctional place for production subordinated to requirements of increasingly heavier and more efficient agricultural machinery. The size of field plots, decrease in the area of permanent grasslands, chemisation and intensification of agricultural production reached its apogee in the 1980s. The negative influence of socialist agriculture on the landscape led to official reports on the state of the environment showing early after 1990 drastic statistics exemplifying the extent of the clearing and liquidation of scattered greenery from the agricultural landscape including 4.000 km of lines of wood vegetation, 3.600 ha of scattered greenery, 49.000 km of balks and 158.000 km of field roads that had been removed from the Czech rural landscape (Moldan et al., 1990).

On the other hand there are also some changes that had a positive environmental effect such as afforestation and spontaneous successive distribution of shrubland on slopes, a dispersal of tree stands and wetlands along unmaintained streams and on other places not suitable for heavy mechanisation and large-scale agriculture. The removal of field balks and margins, solitary and linear scattered greenery from the agricultural landscape was compensated by the creation of a new wilderness. These sites have become a refuge for endangered plants and animals which were forced away from intensively used agricultural lands. If we compare the decrease in permanent greenery from the fields with its increase in abandoned lands, the result is surprising: the total area of permanent non-forest greenery has doubled in the landscape during the period 1950–1990 (Kubeš, 1994; Lipský, 2005).

The traditional character of the Czech rural landscape with its small-scale mosaic of patches has changed into a large-scale landscape of collective open fields (Lipský, 1995).

On the contrary in southeast Poland, where private ownership and a traditional way of farming remained during the socialist era, the small-scale landscape has been preserved to the present day. This specific regional type of agricultural landscape that was named "Poland Strip Fields" was distinguished as one of 30 significant Pan-European landscape types in the first Pan-European landscape typology (Meeus, 1995). Many Englishmen and Dutchmen, who remember their countries from the 1950s and 1960s, say when they see this Polish landscape: "This is how I remember nature of my childhood. I never thought I would see it, and I found it here, in Poland" (quoted by Szukay, 2009, Nature and landscape protection in Poland, unpublished).

2.5 Present Trends in European Rural Landscapes: Intensification and Extensification

The secondary landscape structure formed by the use of land has changed repeatedly throughout history, depending on political, economic, technological and demographic changes (Rabbinge, van Latesteijn, & Smeets 1996). Agricultural as well as other cultural landscape types are among those that change most rapidly. The transformation of the European agrarian society into an advanced industrial one accelerated after World War II. In recent decades, European agriculture has become increasingly industrialised and more specialised. Thus, traditional rural landscapes, which were the result of the agrarian society, transformed into modern, industrial or even post-industrial landscapes according to Lemaire (2002 in Antrop, 2008).

For most European countries, agriculture is still the most important land-use activity influencing landscape character and biological diversity (Mander & Jongman, 2000). The modernisation of agriculture brings about changes in the landscape. Recent and present developments in the Czech as well as the European rural landscape are characterised by two antagonistic tendencies: intensification and extensification. These different trends can be followed up from the mid Twentieth century. Intensification of food production is a key modern agricultural activity. The use of fertilisers and fossil fuels have made it possible to produce more on less land and this has had – and will continue to have – implications for land use and landscape character. A significant decrease in the area of both arable and agricultural lands in Europe during the last 50–60 years has been accompanied by the generally enormous increase in the intensity of farming on plots that remained for agricultural use, especially in regard to arable lands. Large-scale blocks of arable lands have been regarded only as a monofunctional production space with the aim of maximising agricultural production.

At the same time the process of extensification manifested by marginalisation and abandonment of agricultural lands began to appear in rural landscapes in Europe. In the marginalisation process, land is managed less intensively or it is abandoned. Less intensive use of agricultural lands began to be practiced more with the creation of the EU agricultural policy in the 1980s. In many areas the farming practices associated with landscapes have lost their competiveness. In these areas, typically with a low productivity of soils, land management is at risk (Raes, 2008).

The decrease of anthropic pressure on the landscape is certainly positive from the view of landscape ecology. There are, and in the future certainly will be considerable regional differences between regions of intensive agriculture in the fertile lowlands with primary productive functions on one hand and highlands, mountains and foothills on the other hand. Farmland in these regions being not able to compete in terms of food production can be expected to be released for other land-use forms and functions. Afforestation is the first measure, however it cannot be considered as a universal solution and the only use of land unsuitable for intensive agricultural production. Afforestation and grassing will certainly represent a positive feature in the areas declared as zones of water source protection.

Many small-scale agricultural plots not suitable to modern industrial and market-oriented agriculture were abandoned during the last decades. In some regions, especially in mountains and highlands or in regions of South and North Europe, the process of extensification can be dominant for the whole landscape. In most parts of Europe, however, a total marginalisation is the exception. Marginalisation usually concerns only smaller parts of the land and it can be regarded as a compensation for intensively used arable lands. Processes are mostly a mixture of both intensification and marginalisation (Jongman & Bunce, 2000).

The general trend of recent rural landscape changes is one of polarisation between more intensively and more extensively used land. Equally, intensification and marginalisation increase the polarisation rate of landscapes (Mander & Jongman, 2000). This polarisation means that the current changes are not restricted to the main production areas but all landscapes are affected (Antrop, 2008). In many cases intensification of land use in one area causes marginalisation in other areas (Mander & Jongman, 2000). This development was typical for the Czech countryside during the socialist collective farming period and continued after 1990 under new political and socio-economic conditions (Lipský, 1995, 2005).

2.6 Abandoned Lands and New Wilderness in European Cultural Landscapes

2.6.1 The Origin of the New Wilderness and its Causes

The area of arable as well as total agricultural land had been permanently decreasing during the whole second half of the Twentieth century in our landscape. This development has also been confirmed by statistical data on land use (Bičík & Kupková, 2005; European Environment Agency, 2006), however the real land use and land cover is usually a little different. Maintenance of the rural landscape becomes impossible in some parts whether for technological or economic reasons. Even during the period of socialist agriculture, when a strict law concerning protection and use of agricultural lands was applied and economic aspects were not determinant, some plots and localities not suitable for large-scale agriculture and heavy mechanisation remained as fallow lands. Most abandoned lands were still officially recorded as agricultural land in statistical statements. The area of abandoned lands has been increasing slowly but no official statistics exist, only rough estimates of circa 350.000–400.000 ha in the country. That is approximately 5% of the area of the Czech Republic. Significant regional differences occur among mountains, highlands and fertile lowlands (Lipský, 2005). But it is essential to say that none of the catastrophic forecasts estimating that about half of the area of agricultural land would be left abandoned in the country after 1990 have been fulfilled.

Biotic processes of natural succession and natural stabilisation began on abandoned agricultural lands. Self-seeding trees, shrubs and other seminatural communities began to grow and expand in these localities. They became local centres of biodiversity as refuges for wild species driven away from intensively used agricultural plots. After 50 years of this development we can find many small landscape segments with different successional stages of seminatural vegetation in the Czech rural landscape. Small water stream erosion valleys in the low highlands of Central Bohemia are among typical examples of such development. Natural and seminatural communities originated both in the wet bottom of the valleys along the water stream, where narrow strips of alluvial meadows were previously manually managed, and on relatively steep slopes of the valleys which were formerly used as dry extensive pastures with some low-yielding fruit trees. Whole valleys of small water streams strengthened their biocorridor functions in this way. For many wild species they became a refuge and the only functional biocorridor in the contemporary agricultural landscape.

Two concrete examples from Central Bohemia concisely illustrate the development of the "wet" wilderness in partly abandoned valleys of small water streams.

(a) *Jevanský potok brook* (40 km east of Prague): land use changed on 38% of the alluvial floodplain in the period 1990–2005, chiefly because of abandonment, afforestation and grassing on arable lands. More than 20% of the alluvial plain is now abandoned and covered by a varied mosaic of wet meadows, reed and sedge communities as well as alluvial willow and alder forests in initial succession stages.

(b) *Liběchovka and Pšovka brooks* (50 km north of Prague, total length of investigated valleys 25 km, area 14 km^2): significant land-use changes completely changed the landscape character of both valleys from an open intensively used agricultural landscape to a closed forested landscape scenery (Table 2.2). The land has been rewilding and forest has taken over. This development was started by the transfer of the German population after WWII and accelerated during the subsequent transition to socialist large-scale agriculture. The area of cultivated land dramatically decreased because small-sized agricultural plots on the wet bottom of valleys were not suitable for heavy mechanisation. Completely new wetlands developed in abandoned alluvial floodplains along both water streams during the last 60 years. In 1997 both valleys were declared as one of in total 12 Ramsar Sites (wetlands of international importance) in the Czech Republic.

Table 2.2 Land-use changes in the Liběchovka and Pšovka valleys 1845–2000, as a percentage

Land-use category	1845	1938	2000
Forest and shrub	48	51	70
Permanent grasslands	13	16	15
Arable lands	25	23	3
Total agricultural lands	45	40	18

2.6.2 Terminology and Typology of the New Wilderness

The existence and further development of the so-called "new wilderness" in present European cultural landscapes represents undoubtedly a frequently discussed issue. First of all we should explain the term "wilderness". According to the explanatory dictionary, wilderness is defined as an area of wild uncultivated land, usually far from habitation, but sometimes refers to wild land in an urban area (Webster, 1987).

In the word "wilderness" the emphasis is placed on the objectively existing difference in comparison with a commonly cultivated land. Similar conclusions were made by Míchal (2001), who furthermore defines the term wilderness on the ecosystem level. According to Míchal, the development of the wilderness is not determined from without but by inner movement without any defined goals or time limits. Diverse concepts of wilderness have in common that they have as their basis the things grown fully by oneself (not created by man) and that conform with oneself.

The attribute "new" wilderness shall accentuates the difference in comparison with primary "old" wilderness, represented in Central Europe only by fractional fragments of virgin forests, developing for hundreds and thousands of years without the influence of the man. Old wilderness characterised by climax communities is very rare, endangered and strongly protected in the European cultural landscape. To the contrary new wilderness is characterised by initial and early successional stages of vegetation, not older than approximately 50 years. It is not rare, but expanding, perceived as unwanted and unprotected, of course. New wilderness originates and develops on sites previously used by man. Fallow agricultural lands are considered to be the most extended wilderness in the contemporary landscape of the Czech Republic.

The succession of shrub and forest communities resulting from abandoning agricultural lands completely changed the landscape character in some parts of the country, especially in the above-mentioned erosion valleys of small water streams. It is possible to distinguish different types of new wilderness according to the duration of their existence in the landscape, speed and type of succession, type of communities and site conditions.

According to the former land use, new wilderness can be classified as:

- postagrar (the most common – on abandoned agricultural lands; it can be further divided into wilderness developed on former meadows, orchards, arable lands, gardens etc.);
- postmining (in quarries, sand pits, dumps etc.);
- postindustrial;
- postsettlement.

The great diversity of plant communities under the diverse abiotic conditions is a characteristic feature of the new wilderness: grasslands, steppe and forest steppe vegetation, shrub vegetation, forest vegetation of different species composition, wetlands, reed and sedge vegetation, initial alder and willow alluvial forests etc. The diversity of communities depends significantly on the time the new wilderness has existed, of course. The development of the new wilderness has been too short so far.

There is the danger of a decrease in ecosystem and landscape diversity in the future if climax communities predominate.

Generally speaking, we can distinguish "wet wilderness" existing on wet sites (especially in alluvial plains) and "dry (xeric) wilderness" on xeric sites (steep slopes with rock outcrops).

2.6.3 Importance (and Advocacy) of the Existence of the New Wilderness

As discussed above, human beings are the primary cause of the formation of wilderness in the cultural landscape. But localities where this development takes place are predetermined by natural conditions in the first place. Agricultural lands remain abandoned especially in areas not suitable for modern large-scale technologies involved in agricultural production like steep slopes of valleys and seasonally or permanently wet stands in undrained alluvial plains along water streams. Significant regional differences in distribution of abandoned lands between lowlands and highlands are also firstly caused by natural conditions (Lipský, Kopecký, & Kvapil, 1999).

In contrast with the process of intensification, environmental and landscape ecological consequences of marginalisation and abandonment of agricultural lands are accepted inconsistently even by specialists. While intensification, widely described and analysed in many countries, is evaluated negatively from the landscape ecology point of view, the process of extensification has not been evaluated consistently and uniformly. Some changes are universally welcomed, others may cause conflicts. Changes that are positive in some respects may be negative for other landscape values. A topical problem stems from the risk of elemental abandoning of agricultural land cultivation in marginal regions, which intrinsically promotes the danger of rural region depopulation, breakdown of historical settlement structure, extinction of characteristic features and aesthetic values of the traditional cultural landscape (Jongman, 1996).

Different aspects, both positive and negative, of the process of rewilding and existence of the new wilderness in the contemporary cultural landscape are summarised in Table 2.3.

Igor Míchal (2002) noted four leading motives for letting the process of rewilding take its course and for protecting such new wilderness in the present cultural landscape of the Central Europe:

- *Ecological* (this concerns knowledge of natural processes especially succession of communities, biogeochemical cycles, trophic chains, ecological stability, biodiversity, island biogeography etc.);
- *Utilisation-functional* (importance of nature for man, wise use, caring management and servanthood stewardship);
- *Ethical* (positive ethical relations to wilderness resulting from ideal integration of man and nature, appreciation of inner values of nature and wildlife);
- *Psychological-emotional* (wilderness as the opposite to a managed landscape, positive emotional relations to natural elements).

Table 2.3 Positive and negative aspects of the new wilderness in the cultural landscape

Positive aspects (+)	Negative aspects (–)
Compensation of intensively used arable lands Space for natural processes, especially succession of natural communities Increase in ecological stability of the landscape Increase in the area of ecologically stable landscape segments like forests, shrub, steppe and wetland communities (Temporary?) increase in ecosystem and species biodiversity Strengthening of biocorridor functions of alluvial plains and river valleys Origin of biocentres and refugia for many plant and animal species Increase in vegetation index with positive climatic consequences Water retention in the landscape No damage during floods	Some native species are endangered by the change Wildlife dependent upon agricultural practices are threatened Decrease in ecosystem and species biodiversity Possible spread of invasive species Change in landscape character Traditional regional rural landscape types are under threat and vanish Worse passability of the landscape (for man only) Worsens people's landscape perception (especially for farmers, owners, stakeholders)

2.7 Conclusions

Landscape is becoming an integrative concept. There is a growing need for transdisciplinary research (Naveh, 2000; Antrop, 2008). Landscape changes represent a significant issue in contemporary Europe. Two aspects can be recognised: traditional cultural landscapes become lost and disturbed, and the growing speed and magnitude of ongoing changes (Antrop, 2008). Landscape changes have always taken place, but today this is too often coupled with loss of character. Today's fast changing society and environment result in the creation of completely new landscapes and in rapid deterioration of traditional ones, which is considered a threat to quality and values. The richness and diversity of rural landscapes in Europe is still regarded as a distinctive feature and an integral part of the natural and cultural heritage of the continent (Meeus, 1995). But this heritage is now endangered by the processes of both intensification and extensification. As regional cultural landscape types vanished during the last century some new ones appeared like semiurban or hybrid urban, recreational, post-industrial and post-agrar types of landscapes. It is not possible to say that traditional landscapes are better or worse than contemporary landscapes: the main difference is in our attitude to the environment. There will always be a landscape, but what landscape? This is a new question (Antrop, 2008).

What is undisputed, the changes in land use and landscape structure have many relevant ecological, environmental and even societal consequences. Among 203 threatened habitats in EU countries, 132 are potentially influenced by intensification and 32 by abandonment of human activities (Mander & Jongman, 2000). The assessment of changes in the landscape and of interventions by man into the landscape does not mean a precarious rejection of changes but an evaluation of whether and how the changes comply with or counteract natural processes, whether they affect the landscape's ecological stability and biodiversity negatively, whether they endanger landscape values and exceed its carrying capacity etc. (Lipský, 2000). The focus of research in landscape ecology needs to be on how landscape dynamics interacts with species tolerances in time and space (Dunn, Sharpe, Guntenspergen, Stearns, & Yang, 1991).

Present trends in developments and different aspects of the existence of the new wilderness in the cultural landscape are now a matter of discussion among scientists, landscape planners, managers and stakeholders. There are widely varying opinions from specialists as well as stakeholders concerning current landscape changes, especially concerning the abandonment of agricultural lands. The origin of the new wilderness causes a serious dilemma for nature and landscape conservation: whether to resist the natural processes of succession and ecological stabilisation of the landscape in favour of the protection of some species and a traditional landscape character or let natural processes take their course?

As every cultural landscape is a mirror of the state and development of society, man carries a great responsibility for the state of the landscape and its functions and values as well as having a possibility to improve them. Transition from the industrial to the post-industrial global information age – this is a crucial period of great dangers but also of great opportunities in which we are confronted with the choice between further evolution on the Earth or its final extinction (Naveh, 2000). One of the main problems in contemporary landscape planning and management is the high number of actors that have territorial competence. In a definitive effect, the land owner (private or public) is the only one who can make real, material and tangible changes (Antrop, 2008).

Acknowledgments The paper was prepared with support of the research plan of the Czech Ministry of Education (No. MSM 0021620831) *Geographical Systems and Risk Processes in the Context of Global Changes and the European Integration as well as the research project of the Czech Ministry of Education (2B06013) Implementation of the European Landscape Convention in Intensively Utilised Agricultural Regions with Significant Signs of Historical Landscape Design Activities – A Pilot Study Nové Dvory-Kačina.*

References

Antrop, M. (2008). Landscapes at risk: About change in the European landscapes. In: P. Dostál (Ed.), *Evolution of geographical systems and risk processes in the global context* (pp. 57–79). Prague: Charles University.

Aspinall, R. (2006). Editorial. *Journal of Land Use Science, 1*(1), 1–4.

Bičík, I. (1998). Land use in the Czech republic 1845–1948–1990. In: *Methodology, interpretation, contexts*. Geographica-Suppl. Prague: Acta Universitatis Carolinae.

Bičík, I. Jeleček, L. (2003). Long term research of LUCC in Czechia 1845–2000. In: L. Jeleček, et al. (Eds.), *Dealing with Diversity* (pp. 224–231). Prague: Charles University.

Bičík, I., & Kupková, L. (2005). Long-term changes in landscape use in czechia. *Historická Geografie (Historical Geography), 1*(33), 346–366 (in Czech).

Björklund, K. (1996). *The transitional landscape*. Man and Nature Working Paper 83. Odense: Odense University.
Brassley, P. (1997). *Agricultural technology and the ephemeral landscape*. Man and Nature Working Paper 94. Odense: Odense University.
Council of Europe. (2000). *European landscape convention and its explanatory report*. Strasbourg.
Dunn, C. P., Sharpe, D. M., Guntenspergen, G. R., Stearns, F., & Yang, Z. (1991). Methods for analyzing temporal changes in landscape pattern. In: M. G. Turner & R. H. Gardner (Eds.), In *Quantitative methods in landscape ecology. The analysis and interpretation of landscape heterogeneity. Vol. 82. Ecological Studies* (pp. 173–198). Berlin, Heidelberg, New York: Springer.
European Environment Agency. (2006). *Land accounts for Europe 1990–2000. Towards integrated land and ecosystem accounting*, EEA Report No 11/2006. Copenhagen.
Forman, R. T. T., & Godron, M. (1986). *Landscape ecology*, New York: John Wiley & Sons.
Gabrovec, M., & Petek, F. (2003). Changes in land use in the border zone of Slovenia and Austria – the case of the svečinske gorice hills. In: L. Jeleček, et al. (Eds.), *Dealing with diversity*, (pp. 237–241). Prague: Charles University.
Himyiama, Y., Mather A., Bičík, I., & Milanova, E. V. (Eds.). (2005). Land use/cover changes in selected regions of the world. In: *IGU study group on land use/cover change*. Sapporo: Hokkaido University of Education.
Jongman, R. H. G. (Ed.). (1996). *Ecological and landscape consequences of land use change in Europe*. Tilburg: ECNC Publication Series on Man and Nature.
Jongman, R. H. G., & Bunce, R. G. H. (2000). Landscape classification, scales and biodiversity in Europe. In: Ü. Mander & R. H. G. Jongman (Eds.), *Consequences of land use changes in Europe. Vol. 5. Advances in ecological sciences* (pp. 11–38). Southampton: WIT Press.
Kolejka, J. (2002). Czech experience with LUCC research. In: I. Bičík, et al. (Eds.), *Land use/land cover changes in the period of globalization* (pp. 144–152). Prague: Charles University.
Krausmann, F. (2001). Land use and industrial modernization: an empirical analysis of human influence on the functioning of ecosystems in Austria 1830–1995. *Land Use Policy, 18*(1), 17–26.
Kubeš, J. (1994). Bohemian agricultural landscape and villages, 1950 and 1990: Land use, land cover and other characteristics. *Ekológia/Ecology (Bratislava) 13*(2), 189–198.
Lipský, Z. (1991). Historical development of landscape structure and ecological consequences in the Czech landscape. In: *Practical landscape ecology* (pp. 45–51). Proceedings of the European IALE – Seminar, Vol. II. Roskilde: Roskilde University.
Lipský, Z. (1995). The changing face of the Czech rural landscape. *Landscape and Urban Planning, 31*(1), 39–45.
Lipský, Z. (2000). *Investigation of changes in cultural landscapes*. Prague: Czech Agricultural University, (in Czech).
Lipský, Z. (2005). Landscape changes. *Park-Zahrada-Krajina, 15*(4), 2–6 (in Czech).
Lipský, Z. (2008). Land use and landscape structure changes from the point of view of landscape ecology. In: M. Boltižiar (Ed.), *Implementation of landscape ecology in new and changing conditions*. In Proceedings of the 14th International Symposium on Problems of Landscape Ecological Research (pp. 225–230). Nitra: Institute of Landscape Ecology, Slovak Academy of Sciences.
Lipský, Z., Kopecký, M., & Kvapil, D. (1999). Present land use changes in the Czech cultural landscape. *Ekológia/Ecology (Bratislava), 18*(1), 31–38.
Mander, Ü., & Jongman, R.H.G. (Eds.). (2000). *Consequences of land use changes Vol. 5. Advances in ecological sciences*. Ashurst Lodge, Ashurst, Southampton: Wit Press.
Meeus, J. (1995). Chapter 8. Landscapes. In: P. Bourdeau & D. Stanners (Eds.), *Europeś environment* (pp. 172–189). Copenhagen: The Dobříš Assessment. European Environment Agency.
Moldan, B., et al. (1990). *The Environment of the Czech Republic* Prague: Ministry of Environment and Academia Publishers (in Czech).

Míchal, I. (2001). Fhylogenetically stabilised aspects of landscape perception. In *Tvář naší země. Krajina Domova Vol. 1. Krajina jako přírodní proctor* (pp. 16–21). Prague: Studio JB, (in Czech).

Míchal, I. (2002). Wilderness as a cultural object. In I. Míchal (Ed.), *Tvář naší země. Krajina Domova. Vol. 5. Ochrana krajiny* (pp. 25–30). Prague: Studio JB (in Czech).

Mücher, C.A., Bunce, R. G. H., Jongman, R. H. G., Klijn, J. A., Koomen, A., Metzger, M. J., et al. (2003). *Identification and characterisation of environments and landscapes in Europe.* Alterra-Report No. 832. Wageningen: Alterra.

Naveh, Z. (2000). The contribution of landscape ecology to the sustainable future of post-industrial rural landscapes. In: Ü. Mander & R. H. G. Jongman (Eds.), *Landscape perspectives of land use changes Vol. 6. Advances in ecological sciences* (pp. 191–209). Ashurst Lodge, Ashurst, Southampton: WIT Press.

Olah, B., & Žigrai, F. (2004). The meaning of the time scale transformation of the landscape for its sustainable use. *Ekológia/Ecology (Bratislava), 23* (Suppl 1), 231–243.

Pietrzak, M. (2001). Landscape transformations – major spatial processes. In: K. German & J. Balon (Eds.), *Transformations of the natural environment of Poland and its functioning. Vol. X. The problems of landscape ecology* (pp. 28–36). Kraków: Institute of Geography and Spatial Management of the Jagellonian University, Polish Association for Landscape Ecology (in Polish).

Rabbinge, R., van Latesteijn, H. C., & Smeets, P. J. A. M. (1996). Planning consequences of long term land use scenarios in the European Union. In: R. H. G. Jongman (Ed.), *Ecological and landscape consequences of land use change in Europe*, (pp. 17–35). Tilburg: ECNC Publication Series on Man and Nature 2.

Raes, C. (2008). Landscape: Linking rural development and agriculture. *Newsletter Landscape Europe, 11*(2008), 5.

Wascher, D. M. (Ed.). (2005). *European landscape character areas. Typologies, cartography and indicators for the assessment of sustainable landscapes.* Final Project Report of the ELCAI Project, Alterra-Report no. 1254. , Wageningen: Landscape Europe.

Webster, A. M. (1987). *Ninth collegiate dictionary.* Springfield, MA: Merriam-Webster INC, Publishers.

Zonneveld, I. S. (1995). *Land ecology*, Amsterdam: SPB Academic Publishing.

Chapter 3
Environmental Stressors as an Integrative Approach to Landscape Assessment

Jiří Anděl, Martin Balej, and Tomáš Oršulák

3.1 Stressors and Stress in a Landscape

The genesis of the term stress is closely associated with research in the psychological and biological disciplines (e.g. Shanteau & Dino, 1993). Generally, stress is a difficult concept to define. Early definitions varied in the extent to which they emphasized the responses of the individual, or the situations that caused disruptions of ongoing behaviour and functioning (Evans & Cohen, 1987). Appley and Trumbull (1967), McGrath (1970) and Mason (1975) have summarized several objections to each of these approaches to defining stress. Stress is best considered as a complex rubric reflecting a dynamic, recursive relationship between environmental demands, individual and social resources to cope with those demands, and the individual's appraisal of that relationship (Evans & Cohen, 1987). A stress-inducing factor is called a stressor. Stress is a manifestation of a stressor within a system. Four general types of environmental stressors have been identified in psychological theory: cataclysmic events, stressful life events, daily hassles, and ambient stressors (Baum, Singer, & Baum, 1982; Cambell, 1983; Lazarus & Cohen, 1977.)

Similar to the definition of stress in psychology, we can designate as a stressor any force or system of forces producing pressure, tension or causing deformity that is detrimental to the system it acts upon. In the context of environmental sciences, stress within an environmental system composed of biotic, abiotic and human elements can be defined as any deformity present in the system. Stress (or pressure, strain, disturbing force, obstacle or difficulty) can thus be defined as any stimulus the intensity of which is in excess of the norm (physical, ecological, social or economic). In the normal fluctuation of a phenomenon, stress can be represented by an exceptionally strong/weak intensity or unusual frequency. Individual types of environmental systems may react in varying ways to different stressful stimuli. In the initial phase of stress response, a system operates on the principle of resilience, followed by the phase of resistance. In the final phase the system either breaks down

J. Anděl (✉)
Department of Geography, Jan Evangelista Purkyne University in Ústí nad Labem, České mládeže 8, 40096 Ústí nad Labem, Czech Republic
e-mail: jiri.andel@ujep.cz

J. Anděl et al. (eds.), *Landscape Modelling*, Urban and Landscape Perspectives 8,
DOI 10.1007/978-90-481-3052-8_3, © Springer Science+Business Media B.V. 2010

entirely (i.e. changes its character) or compensates for the stress and continues to function as before. Stress therefore can be compensated for entirely or partially, or it may not be compensated at all and the system breaks down. Some landscape ecologists (cf. Ingegnoli, 2002) claim that if the effect of a stressor is continuous (chronic), this may endanger the general "health" of the landscape.

The Slovak school of landscape ecology devised a theory of environmental stressors (Miklós et al., 2002; Šúriová & Izakovičová, 1995; Izakovičová, Miklós, & Drdoš, 1997). Other landscape ecologists (Ingegnoli, 2002; Lipský, 1998; Antrop, 2000; Erickson, 1999) also employ the terms environmental stressor, stress or anthropogenic pressure in connection with a negative effect on environmental conditions, the pathology of landscape and anthropogenic disturbances.

Within a landscape system, there are of course natural stressors such as natural disturbances (degradation processes, natural radiation, volcanism, seismic activity and seismic processes). Environmental systems are able to a greater or lesser extent to prepare for the effects of these stressors. The other group includes stressors of anthropogenic nature. Primary anthropogenic stressors are defined by Šúriová and Izakovičová (1995) as anthropogenic areas and lines (e.g. built up areas, mining areas or intensively cultivated agricultural areas or transport lines). Secondary anthropogenic stressors are the phenomena that accompany anthropogenic activity (e.g. the volume and character of waste and pollutants produced, the intensity of erosion, noise, etc.) From a geographical perspective, anthropogenic stressors form a continuous system which may be termed a territorial system of environmental stressors. This system is composed of core, planar and linear stressors.

By monitoring the effect of anthropogenic stressors and quantifying the amount of stress within an environmental system we are able to determine the degree of anthropic footprint or the intensity of disturbing influences in the landscape. This analysis of environmental stress also brings a wealth of information about environmental quality (Adamowicz, Swait, Boxall, Louviere, & Williams, 1997; Bastian et al., 2002). The higher the degree of stress, the lower is the resulting environmental quality.

The terms stressor and stress aptly differentiate the agent and the manifestation, the cause and the consequence. Stress is manifested both in the natural and the social subsystem of the environmental system. We can talk of stress in relation to all landscape elements, both natural and social. Stressors comprise both disturbances and anthropogenic land use areas or the accompanying phenomena such as noise, smell, etc. In the social elements of landscape, stress is manifested by the presence of socially pathological phenomena, high crime rate, unemployment, high divorce rate, low percentage of native inhabitants, etc.

The manifestations or consequences of stress within an environmental system, in landscape-shaping processes, often go unnoticed, even though they form a significant force moulding the character of the landscape.

Attempts to quantify the extent of anthropogenic pressure, anthropogenic impact on landscape based on the representation of land-use classes or on the anthropogenic impact on vegetation have been made (Skowronek, Krukowska, Swieca, & Tucki, 2005). In the main, however, these are only partial conceptualisations of the subject. Measuring the negative impact of human activity may be one suitable indicator for decision-making as to the future character of the landscape (Conway & Lathrop, 2005; Nikodemus, Bell, Gríne, & Liepinš, 2005; Pauleit, Ennos, & Golding, 2005; Plantinga & Andrew, 1996).

In this section we will set out to test a new methodological approach to analysing environmental stress. We will apply the approach to eight landscape types and three time periods and collate the resulting data. What was the development of environmental stress in the different types of landscapes and time periods? What were the reasons for this development? Where are the causal factors? How did the individual indicators or clusters of indicators behave? Are the trends in the development of ecological and social stress mutually related or not? Which shows the greater dynamics of change?

3.2 Environmental Stress Accounting

Environmental stress is an intersection of ecological and social stress. We can understand it as the effect of anthropogenic strain in a given time and place. The evaluation of environmental stress includes the evaluation of negative effects on the individual landscape elements (e.g. air, water, soil or social environment). The individual indicators represent corresponding effects of stress. As the indicators are not equally informative (cf. Ritschelová, Machálek, & Koroluk, 2001), they were given different weightings. The different weightings reflect the power of the synergic effect of each indicator (Balej & Anděl, 2008).

Ecological stress (EcoS) includes the negative impact on natural landscape elements (such as the extent of anthropogenic relief forms, pollutants in the air and water bodies, potential threat of water and wind erosion and damage to forest cover etc.) (Chvátalová, 2005; Raška & Oršulák, 2009). Ecological stress is an aggregated value of the individual indicators weighted by the given weighting (Tables 3.1 and 3.2).

Social stress (SocS) comprises those social aspects we consider negative in the social subsystem. These are in particular demographic indicators aimed at population movement (e.g. negative population increase), structural aspects (e.g. unfavourable age and education stratification of the population, economic variables (e.g. high unemployment)) and spatial aspects (e.g. passive migration balance and high migration fluctuation). Social stress is an aggregated value composed of individual indicators weighted by the given weight (Tables 3.1 and 3.2). The calculation of stress is carried out through an evaluation of individual indicators by means of points, as is usual with similar methods. The maximum numerical range of each indicator is divided into quartiles. The numerical values are assigned as follows: low-range quartile (Q1 = 0), below average (Q2 = 1), above average (Q3 = 2) and high (Q4 = 3 points). The points are then multiplied by the corresponding weights (1 or 2).

Environmental stress (EnviS) is calculated as the sum of ecological and social stress. It is not just the combined value that is important, however, but rather the relative proportions of the ecological and social aspects on the overall value. These can reveal the role of the natural and social subsystems in the time period in question.

Table 3.1 Ecological stress indicators

Index	Group	Indicators	Specification	Weight[a]
A_1	Relief and soils degradation	Degree of anthropogenic transformation	Presence of anthropogenic landforms in %	2
A_2		Potential aeolian and water erosion	degree	1
A_3		Dumping places	degree	2
A_4	Air pollution	Air pollution SO_2, NO_x, air dust	mcg/m^{-3}	2
A_5		Local sources of air pollution	tons/km^{-2}	1
A_6	Water quality	Surface water course quality	Quality factor	2
A_7	Biota	Forestal air pollution zones	prevailing category A, B, C, D, E, F	1
A_8	Other stress	Ecological stability index	Ratio of relatively stable and unstable land	1
A_9		Noise and emission	Intensity and frequency of traffic	2
A_{10}		Barriers	Length of artificial-transportation ways km/km^2	2

[a]According to the assessment made by a team of international experts.

Table 3.2 Social stress indicators

Index	Group	Indicators	Specification	Weight[a]
B_1	Population change	Natality	Average 5 years	1
B_2		Natural increase	Average 5 years	1
B_3		Index of vitality	Preproductive/postproductive	2
B_4	Family relation	Divorce rate	Average 5 years	1
B_5		Incomplete families	In total	2
B_6	Economic relation	Index of education	University/elementary	2
B_7		Unemployment	Average 5 years	2
B_8	Spatial movement-lability	Natives	in %	2
B_9		Migrational balance	Average 5 years	2
B_{10}		Migrational change	Average 5 years	1

[a]According to the assessment made by a team of international experts.

We have tested this methodological approach in the region in question in the period spanning the second half of the twentieth century, which we divided into five time intervals. This period has a number of valid data sources available. Around 1950 the final phase of the industrial period in the Czech Republic had begun – the so-called totalitarian phase. The influence of anthropogenic stressors intensified significantly during this period. The following two time intervals (beginning in 1970 and 1980, respectively) point to the dynamic changes in the environmental system during the totalitarian period (communism). 1990 saw the beginning of the post-industrial period, while 2005 represents the start of a phase in which the influence of stressors in the post-industrial period was attenuated. Overall, we are thus focusing on the final phase of the industrial society in the Czech Republic (the totalitarian period) and on the post-industrial phase, where fundamental changes in ecological and social stress in our study regions can be documented.

In spatial terms, the methodology was tested on typologically differing study regions (Fig. 3.1). Four specific pairs were selected, representing varying geographical types of spatial units at a choric level (Balej et al., 2004): border region (Petrovice) vs. inner region (Třebenice), periphery (Verneřice) vs. core area (Benešov), labile region (Bílinsko) vs. stable region (Libčevesko) and mountainous region (Vejprtsko) vs. lowland region (Klášterecko). However, the stress monitored in the regions indicates the influence of stressors originating not only within those regions but also outside them. The negative effect of stressors of course does not respect administrative boundaries. The influence of larger (pan-regional or province-wide) stressors outside the borders of the study areas thus manifests itself.

All eight study regions lie within the Ústí Region (5.335 km^2, 6.8% of the area of the Czech Republic), which is a region marked by the highest levels of environmental stress in the whole of the country as well as the presence of the greatest

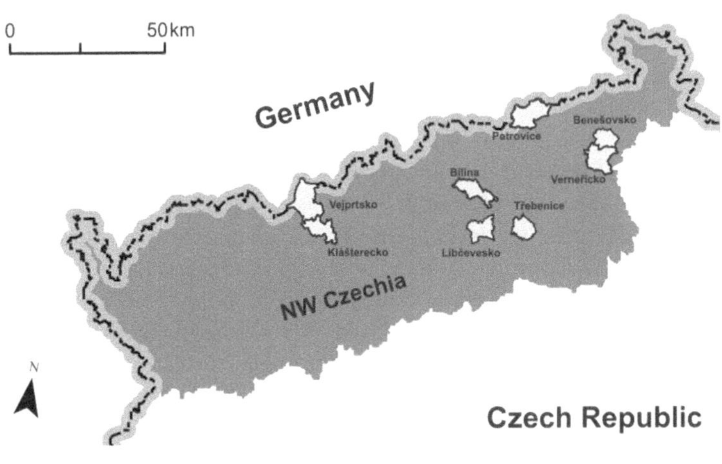

Fig. 3.1 Geographical position of the north-west of the Czech Republic

stressors. The Ústí region (NUTS 4) lies on the border with Germany (Saxony) and its land use is divided into the following categories: farmland greater than 50%, forest areas 30% and bodies of water 2%. As of 2008 its population was 835 thousand inhabitants, 8% of the population of the Czech Republic as a whole. Population density stands at 158 inhabitants per km^2, which is 25% higher than the average for the country. The region is one of the most heavily urbanised areas in the Czech Republic, with 80.5% of its inhabitants living in towns. Unemployment is a significant problem in the region and is still running at the highest levels in the country, despite its reduction in recent years below the 10% level.

In former times the Ústí Region was one of the richest parts of Bohemia, renowned for its high quality agricultural activities. Ore mining for copper and tin developed in the Krušné hory (Ore Mountains), which form a natural boundary between the Czech lands and Germany, while in the foothills of those same mountains open-cast mining for brown coal began as early as the Sixteenth century. Today the Ústí region is one of the most heavily industrialised in the country, and is dominated by the large-scale mining of brown coal and the subsequent production of electrical power. The region is a power/fuel base that also hosts a high proportion of the chemical and food-processing industries.

The Ústí region is an ideal study region on which to test the methodological approach of environmental stress accounting. In the past it has been a barometer that revealed the fundamental changes in the development of the Czech lands, of which it was also frequently at the forefront. This is true both of the processes of industrialisation and urbanisation, which were closely tied with the same processes in Saxony in the mid-Eighteenth century, and of questions linked with ecological problems (Balej et al., 2004).

3.3 Results from Case Studies

Ecological stress rose dynamically during the communist regime (1950–1990) – by almost twofold (Fig. 3.2). While in the two decades between 1950 and 1970 ecological stress increased by 53%, the increasing stress dynamics meant that the same increase was then registered in the course of one decade only, between 1980 and 1990. The region of Ústí nad Labem was then on the brink of ecological catastrophe. Following the political sea change in 1989 and thanks to the implementation of environmental measures, the level of EcoS was brought down by 18.6% between 1990 and 2005. Spatial differentiation of ecological stress was on the rise during the communist period, with the range of values almost doubling (14–25 points) between 1950 and 1990. In the post-industrial period, on the other hand, the degree of ecological stress shows a stagnating trend, with the spatial differentiation of stress also decreasing.

The development of individual indicator clusters reveals a marked differentiation in ecological stress (see Fig. 3.2). Air pollution indicators showed the most dramatic decrease (as a result of the desulphurisation of coal-driven power plants or their

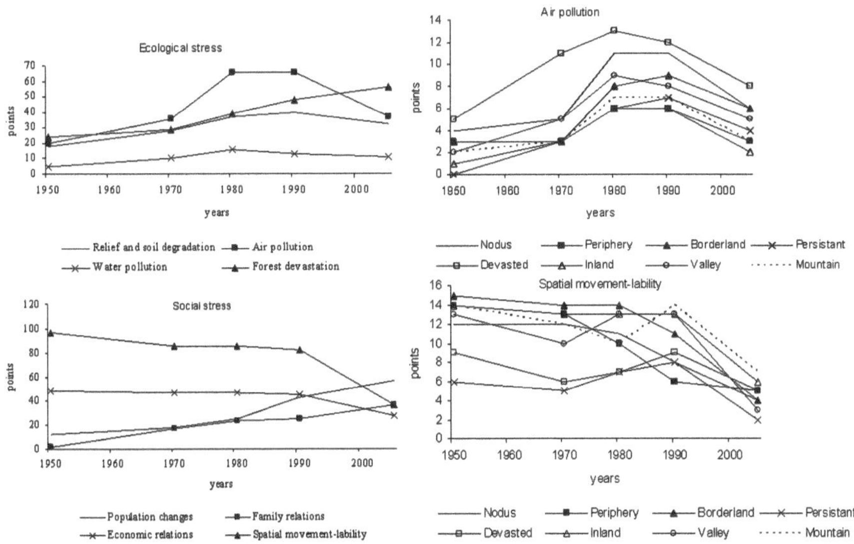

Fig. 3.2 The development of environmental (ecological and social) stress in the study areas in the Ústí nad Labem region compared with selected indicators

decommissioning, the introduction of fluid boilers, the shift to high-grade fuels – gassification, and the shutting down of operations producing excessive air pollution). The negative impact on water courses also decreased (as a result of the creation of wastewater treatment stations in settlements and industrial plants and a monitoring and sanctions policy towards wastewater producers). The degradation of relief and soils also went down very slightly. However, the situation as reflected by the biota indicators remained grave. The intensification of transport resulted in increasing levels of noise pollution and air pollution and also in increasing fragmentation (barrierisation) of landscape.

The degree of ecological stress is strongly influenced by geographic location. The most polluted localities were to be found in the low lying or valley or basin-shaped localities of the Ústí region which contains the most potent ecological stressors in the whole of the Czech Republic. A high degree of pollution was also found in the Ore Mountains during the Communist period due to the high smoke-stacks of the coal-powered power stations. The predominant north-westerly wind would carry the pollutants great distances and to high altitudes. The most pollutant-free air was – and still is – to be found in the peripheral localities "in the shade" of the České Středohoří (the Verneřice and Libčeves regions).

Social stress was at its highest in the post-war period as a consequence of the exodus of the German population and insufficient immigration (Fig. 3.2). The whole settlement structure of the Czech-German borderland was disturbed (Jeřábek et al., 2004). Small settlements in peripheral locations frequently disappeared altogether. The population of newcomers on the whole lacked any bonds to the territory, which resulted in markedly unstable tendencies (from the point of view of migration), namely high migration turnover. The core towns below the Ore Mountains expanded the most dynamically, to some extent at the expense of the surrounding rural areas. Because of the frequently conflicting tendencies of the individual aspects of social stress, its levels changed comparatively moderately and gradually between 1950 and 1970. The level of social stress did not rise until the period of 1970–1990. The post-1990 period is characterised by a sharp drop in social stress, back to the level of 1950. Analogously to ecological stress, social stress also shows an increasing spatial differentiation – which almost doubled in 1950–1990, as it did for ecological stress.

The set of indicators designated as spatial mobility indicators showed a positive trend. The trend confirms increasing population stability and its gradual "merging" with the landscape that the people inhabit. A contrasting tendency is revealed in the "population change" set of indicators as well as in the set relating to "family ties". In the post-industrial period the demographic characteristics follow the trends prevailing in Europe: the birth rate is decreasing (from 20 to 10%), while mean life expectancy is on the rise, as is the percentage of children born out of wedlock. From an economic perspective, there is rising unemployment (above 10%), which was officially non-existent during the communist period for ideological reasons. The rise of the education index is a positive trend, as are the stabilising tendencies in the settlement system – with the percentage of natives rising, migration turnover decreasing and the migration balance registering positive values. Spatial mobility of the population shows marked differentiation. The peripheral locations of the Ústí region show dynamic, positive changes, whereas the core locations do not register marked changes until after 1990. There is an unequivocal trend from polarisation to levelling.

Using the Pearson correlation coefficient we can ensure the correct selection of environmental stress indicators, or even the existence of potential statistical relations between individual indicators or sets of indicators. (Tables 3.3 and 3.4). There are comparatively strong correlations (bonds) between the indicator sets quantifying ecological stress. This is because of the high degree of interdependence of the individual landscape elements, the high degree of interconnectedness within the ecological subsystem. Disturbed relief has a negative impact on water, biota and consequently air quality. Industrial operations, especially mining and the related power industries, impact on all the components of the ecological subsystem in the Ústí region. In the case of biota, there is, logically, a correlation of medium closeness ($r > 0.3$), as other factors come into play (e.g. agricultural and forestry activities). Of the individual indicators, the closest pair correlations can be detected in the degree of anthropogenic relief transformation with legacy environmental issues ($r = 0.708$), in the high quality of surface water ($r = 0.658$), and in the indicator measuring the damage disturbance to forest cover ($r = 0.760$).The disturbance of forest cover is closely correlated with the anthropogenic transformation of the relief, the legacy environmental issues ($r = 0.731$), the quality of surface waters ($r = 0.709$). Surprisingly, the correlation with air pollution is only moderately close ($r = 0.384$ and 0.406, respectively). It becomes apparent that disturbance of forest cover tends to linger and remains high long after the air quality has improved. The

Table 3.3 The Pearson correlation coefficient applied to individual ecological (A1 and A2) stress indicators to determine their mutual correlations

(A1) Ecological indicators

	A_1	A_2	A_3	A_4	A_5	A_6	A_7	A_8	A_9	A_{10}
A_1		−0.324	0.708	0.341	0.158	0.658	0.760	0.060	0.453	0.250
A_2			−0.033	−0.071	0.503	0.113	0.245	−0.279	−0.219	0.000
A_3				0.665	0.190	0.715	0.731	0.203	0.396	0.357
A_4					0.302	0.419	0.384	0.436	0.171	0.004
A_5						0.129	0.406	−0.365	0.060	−0.288
A_6							0.709	0.441	−0.030	0.555
A_7								−0.032	0.336	0.237
A_8									−0.389	0.211
A_9										0.199
A_{10}										

(A2) Ecological indicators

	Relief and soil degradation	Air pollution	Water pollution	Forest devastation
Relief and soil degradation		0.776	0.915	0.405
Air pollution			0.685	0.352
Water pollution				0.316
Forest devastation				

relaxation time is long, and forests therefore respond slowly. This is further confirmed by the low correlations compared to other indicator sets (polluted air and water, degradation of relief and soil) which are closely interconnected.

Compared to ecological stress, the correlations in the case of social stress are far more complex and difficult to interpret. This can be attributed to the far greater complexity and developmental entanglement (or almost contingency) within the social subsystem. We can detect a medium-close correlation between individual sets of social stress indicators; in some cases there is even a negative correlation. The spatial mobility of the inhabitants (indicating lability or stability of the study area) is logically related to the economic aspects, as an economically thriving locality attracts immigration ($r = 0.664$). On the other hand, less economically viable locations may become depopulated ($r = 0.410$) or suffer from disturbed family bonds ($r = 0.521$). The psychological effect (loss of illusions) is likely to have an important role in this respect. Of the individual social stress indicators, a relatively close pair correlation with the ratio of natives is detectable. In localities with stable population (high ratio of natives) we can detect a low vital index ($r = 0.591$) and low birth rate ($r = 0.654$), but comparatively favourable education structure ($r = -0.591$) and undisturbed family bonds ($r = 0.735$). In case of the vital index, closer correlations with other social stress indicators are detectable. The vital index has a logical

Table 3.4 The Pearson correlation coefficient applied to individual social (B1 and B2) stress indicators to determine their mutual correlations

(B1) Social indicators

	B_1	B_2	B_3	B_4	B_5	B_6	B_7	B_8	B_9	B_{10}
B_1		0.395	0.432	0.052	0.754	–0.717	...	−0.654	−0.129	−0.429
B_2			0.580	0.101	0.056	–0.121	...	−0.367	0.062	−0.274
B_3				0.355	0.004	–0.236	...	−0.591	0.386	−0.558
B_4					0.125	–0.053	...	−0.262	0.504	−0.562
B_5						–0.666	...	−0.400	−0.253	−0.319
B_6								0.574	0.290	0.391
B_7								...	...	...
B_8									−0.063	0.735
B_9										−0.342
B_{10}										

(B2) Social indicators

	Population changes	Family relations	Economic relations	Spatial movement-lability
Population changes		0.347	–0.410	–0.517
Family relations			–0.521	–0.444
Economic relations				0.664
Spatial movement-lability				

correlation with natural population increase ($r = 0.580$) and indirectly to the ratio of natives and migration turnover ($r = –0.558$). This medium-close bond can be interpreted to mean that areas with more post-productive inhabitants tend to be more labile in terms of migration. This is confirmed by the new migration trends showing increased migration in older populations, in particular due to the quality of the environment.

3.4 Environmental Stress Accounting and Landscape Studies: Evaluation and Prospects

The development of the relationship between man and the landscape may be classified in general terms into different historical periods (Agnew, Livingstone, & Rogers, 1996; Hampl, 1998). In the pre-industrial period, the settlement structure and economic activities were shaped predominantly by natural determinants. Employment in the primary sector was dominant (agriculture in the lowlands, metal ore mining and forestry in the mountainous areas) and the dynamics of development was low. The population was distributed fairly evenly and individual settlements did not differ greatly in terms of size. The settlement structure was homogeneous. In the industrial period the secondary sector evolved dynamically. The influence of natural determinants weakened as socioeconomic factors gradually came to the foreground. The ecological and human subsystems often competed with one another. The urbanisation process, which is associated with high spatial mobility, came to the fore. Overall, environmental stress rose.

The totalitarian period (the Nazi occupation and the Communist period, 1938–1989) represents the final phase of the industrial society. In this period the development of the country diverges from the natural trajectory of western Europe. The natural development of the Ústí region was fundamentally disrupted. The beginning of the disruptive process was the expulsion of the German element in the wake of World War II. The disappearance and destruction of a great number of settlements, communications and landscape landmarks was to follow. Waves of new settlers from the inland parts of Czechoslovakia moved to the region. These new settlers on the whole lacked any historical bonds to the region. The centrally directed economy permeated to all areas of society. In the basins below the Ore Mountains large-scale lignite mining expanded and heavy industry (power and chemical industries in particular) followed suit. Following the waves of collectivisation (nationalisation of agricultural land) intensive agricultural primary production dominated in the fertile locations. Environmental stress rose sharply during the totalitarian period. The powerful impact of supra-regional stressors (air pollutants) disrupted all elements of the ecological subsystem. In the region of Ústí alone, 98 settlements with a total population of 110 thousand inhabitants were destroyed. The social subsystem abounded in negative traits. Ethnic heterogeneity was comparatively high, the education structure unfavourable, and the rate of sociopathic phenomena comparatively high. Family ties became increasingly disrupted, and the percentage of incomplete families rose. Towards the end of this period there was a dynamic spike in emigration in consequence of and as a response to the alarming state of the environment.

The transformation (post-industrial period, 1990–2005) saw the intensive development of communication and information activities, as well as of the tertiary sector (services and tourist industry). It is a period marked by a sharp decrease in stress and a return to the state of things before the beginning of the totalitarian period. Between 1995 and 1997 in the region of Ústí alone, 20 power-plant units were desulphurised, and between 1991 and 1998 1,190 MW power units were decommissioned and the technology of fluid boilers was introduced to many blocks. Thanks to these measures, SO_2 levels dropped by 92% compared to the beginning of the 1990s; particulate emissions by 95%, nitrogen oxides by 50% and carbon monoxide by 77%. The dynamic rise in ecological stress up to 1990 was related both to air pollution and to the increasing degradation of forest cover, the rising extent of anthropogenic transformation of soil and decreasing quality of surface water courses. Conversely, a more dynamic decrease in ecological stress after 1990 in all the study areas was prevented by the stressful effect of intensifying transportation, and consequently rising noise pollution and traffic emissions, rising fragmentation, steady degree of anthropogenic transformation of relief and land, as well as the static condition of forests. The amplitude of ecological stress values is as follows during the monitored periods: The Třebenice region (more than 300%), the Petrovice region (almost 300%), and the Bilina region (more than 200%). As for the settlement structure, depopulating development tendencies came to the fore in the large settlement areas. The number of inhabitants decreased at the expense of localities in their periphery (Hampl, 2001). The changes in geopolitical conditions brought about the development of heretofore peripheral locations at the German border. For many of these localities, the tourist industry (recreational industry) was also the chief agent of economical and social restoration of the locality. Social stress also decreased sharply. The identification of man with the landscape became stronger.

On the basis of the results obtained by applying the methodological approach analysing the development of ecological, social and environmental stress alongside with geographic factors, we may attempt the following simplified typology of the Ústí region:

- type 1: Regionally exposed areas with a predominantly mining and urban landscape function and high ecological and social stress, where the ecological stress strongly prevails over the social stress (the basin areas below the Ore Mountains);
- type 2: semi-peripheral areas with recreational function, with a low degree of stress and a predominance of social stress over ecological stress (the plateaus of the Ore Mountains and the České Středohoří); mostly marginal, abandoned localities, permanently underpopulated after the expulsion of the German element, weakened historical bonds to the landscape;
- type 3: peripheral areas with predominantly agricultural landscape function, where ecological and social stresses are equivalent; areas under intensive agricultural cultivation, with quality soils for plant production.

When researching the development of landscape and landscape change, the quality of the individual compositional elements must not be overlooked. In the case of the Czech-German borderland we have a territory which was for more than four decades negatively shaped and modified by human influence. This stressful anthropogenic imprint is so essential to understanding the region that any attempt at monitoring and interpretation of the development of the landscape without taking this influence into account would be simply erroneous. The objective of our methodical approach is not only to monitor the impact of "endemic" anthropogenic stressors over time in the study areas, for this would have left some of the negative impact manifested in the areas under discussion unaccounted for. Many of the negative effects are informed by stressors located far away from the borders of the study areas and their influence is transferred into these areas.

Acknowledgements The implemented study is one of the partial outputs of the research project of the Ministry of Labour and Social Affairs of the Czech Republic (No 1J 008/04-DP1). The authors are grateful for the support.

References

Adamowicz, W., Swait, J. Boxall, P., Louviere, J., & Williams, M. (1997). Perceptions versus objective measures of environmental quality in combined revealed and stated preference models of environmental valuation. *Journal of Environmental Economics and Management 32*(1), 65–84.

Agnew, J., Livingstone, D., & Rogers, A. (1996). *Human geography – An essential anthology*. Oxford: Blackwell Publishing.

Antrop, M. (2000). Changing patterns in the urbanized countryside of Western Europe. *Landscape Ecology 15*(3), 257–270.

Appley, M., & Trumbull, R. (Eds.). (1967). *Psychological stress*. New York: Appleton-Century-Crofts.

Balej, M., et al. (Ed.). (2004). *Eastern Ore mts. – Geographical assessment of peripheral area* (pp. 252–259). Ústí nad Labem: Acta Universitatis Purkynianae (in Czech).

Balej, M., & Anděl, J. (2008). Land use changes and environmental stress accounting (case study from northwestern part of the Czech-German borderland). *Journal of Geography and Regional Planning, 1*(5), 97–109.

Bastian, O., et al. (2002). Landscape analysis, synthesis, and diagnosis. In O. Bastian & Ü. Steinhardt (Eds.), *Development and perspectives of landscape ecology* (pp. 113–168). Dordrecht: Kluwer Academic Publishers.

Baum, A., Singer, J. E., & Baum, C. (1982). Stress and the environment. In G. W. Evans (Ed.), *Environmental stress* (pp. 15–44). New York: Cambridge University Press.

Cambell, J. (1983). Ambient stressors. *Environment and Behavior, 15*(3), 355–380.

Chvátalová, A. (2005). Physicogeographical approaches to the research on a cultural landscape. In M. Balej & J. Anděl (Eds.), *Complex geographical research on a landscape I* (pp. 69–74). Ústí nad Labem: MINO (in Czech).

Conway, T. M., & Lathrop, R. G. (2005). Alternative land use regulations and environmental impacts: assessing future land use in an urbanizing watershed. *Landscape and Urban Planning 71*(1), 1–15.

Erickson, D. L. (1999). Rural land use and land cover change. *Land Use Policy 12*(3), 223–236.

Evans, G. W, & Cohen, S. (1987). Environmental stress. In D. Stokols & I. Altman (Eds.), *Handbook of environmental psychology* (pp. 571–601). New York: John Wiley & Sons.

Hampl, M. (1998). *Reality, society and geographical organisation. Finding an integral order*. Prague: Charles University (in Czech).

Hampl, M. (Ed.). (2001). *Regional development: Typical features of Czech transformations, European integration and general theory*. Prague: Charles University (in Czech).

Ingegnoli, V. (2002). *Landscape ecology: A widening foundation. A holistic unifying approach*. Berlin, Heidelberg, New York: Springer.

Izakovičová, Z., Miklós, L., & Drdoš, J. (1997). *Landscape ecological conditions of sustainable development*. Bratislava: VEDA (in Slovac).

Jeřábek, M. Dokoupil, J., Havlíček, T., Chromý, P., Marada, M., Rumpel, P., et al. (2004). *Czech borderland: A barrier or an area of internediation?* Praha: Academia (in Czech).

Lazarus, R. S., & Cohen, J. (1977). Environmental stress. In J. Wohlwill & I. Altman (Eds.), *Human behavior and environment*. New York: Plenum.

Lipský, Z. (1998). *Landscape ecology for students of geographical branches*. Prague: Karolinum (in Czech).

Mason, J.W. (1975). A historical review of the stress field (Pts. 1, 2). *Journal of Human Stress, 1*(1), 6–12.

McGrath, J. (1970). Major methodological issues. In J. McGrath (Ed.), *Social and psychological factors in stress* (pp. 41–57). New York: Holt.

Miklós, L., et al. (2002). Landscape atlas of Slovac Republic. Banská Bystrica: MŽP SR, Bratislava and SAŽP (in Slovac).

Nikodemus, O., Bell, S., Gríne, I., & Liepinš, I. (2005). The impact of economic, social and political factors on the landscape structure of the vidzeme upland in Latvia. *Landscape and Urban Planning 70*(1), 57–67.

Pauleit, S., Ennos, R., & Golding, Y. (2005). Modeling the environmental impacts of urban land use and land cover change – A Study in Merseyside, UK. *Landscape and Urban Planning, 71*(2–4), 295–310.

Plantinga, A. J., & Andrew, J. (1996). The effect of agricultural policies on land use and environmental duality. *American Journal of Agricultural Economics, 78*(4), 1082–1091.

Raška, P., & Oršulák, T. (2009). An overview of natural risks in the ustecko region. In J. Anděl, et al. (Ed.), *Complex geographical research on a landscape II* (pp. 93–112). Ústí nad Labem: MINO (in Czech).

Ritschelová, I., Machálek, P., & Koroluk, R. (2001). Integrated presentation of the economic and environmental information for the Czech Republic, 1992–1996/1997. *International Journal of Environment and Pollution, 15*(6), 636–694.

Shanteau, J., & Dino, G. A. (1993). Environmental stressor effects on creativity and decision making. In O. Svenson & A. J. Maule (Eds.), *Time Pressure and stress in human judgment and decision making* (pp. 293–308). New York: Plenum Press.

Skowronek, E., Krukowska, Z., Swieca, A., & Tucki, A. (2005). The evolution of rural landscapes in mid-eastern Poland as exemplified by selected villages. *Landscape and Urban Planning 70*(1–2), 3–8.

Šúriová, N., & Izakovičová, Z. (1995). Territorial system of anthropogenic stress factors in landscape ecological planning. *Ekológia (Bratislava), 14*(2), 181–189.

Part III
Between Landscapes and Multi-Scale Regions

What factors predispose the position of regions in international communities? How does public opinion in the integrated regions differ? What are the reasons for and effects of cross-border cooperation in integrated Europe? How can we intensify and optimise the methods of cross-border cooperation? What are the differences in land-use structure in cross-border areas of the former Iron Curtain? How do the land-use changes relate to functions of and processes in a landscape? What is the role of peripheral areas in regional European landscapes?

Chapter 4
Environment and Regional Cohesion in the Enlarged European Union – Differences in Public Opinion

Petr Dostál

4.1 Issues of Environmental, Regional and Cohesion Policies

This chapter presents analyses of trends of current public opinion on issues of environmental policies and regional and cohesion policies across 27 polities of the European Union (EU). Public opinion and mass interest articulations of national electorates are central to studies on EU policies, because they form an important feedback that often implies barrier effects on policy-making and decision-making of governing political elites of the democratic countries concerned. Economic and social transformations associated with development of current post-industrial societies have resulted in the EU in new challenges for environmental policies and regional and cohesion policies.

Environmental policy-making was a latecomer to the policy agenda of European integration and has gained gradually in importance since the 1970s. The construction of the European Economic Community (EEC) was primarily driven since the 1957 Treaty of Rome by the quantitative considerations of building the common market and paying little attention to its qualitative aspects (McCornick, 2001). In 1987, the Single European Act confirmed that environmental management was one of the formal policy goals of European integration. The environment is now one of the primary policy interests of the EU. The Single European Act instituted an explicit legal basis upon which environmental protection could operate. However, the internal market measures were to be decided through qualified majority voting in the Council of Ministers, those concerning environmental protection required the unanimity of all Member States. The 1993 Maastricht Treaty listed the environment as a key policy goal of the EU and extended qualified majority voting to environmental policy-making and also strengthened the role of the European Parliament in this sector of policy-making (Dinan, 2005).

P. Dostál (✉)
Department of Social Geography and Regional Development, Charles University in Prague, Albertov 6, 128 43 Praha 2, Czech Republic
e-mail: dostal@natur.cuni.cz

J. Anděl et al. (eds.), *Landscape Modelling*, Urban and Landscape Perspectives 8,
DOI 10.1007/978-90-481-3052-8_4,

Also regional and cohesion policy-making is one of the primary policy interests of the EU and it is concerned with the reduction of economic and social disparities between richer and poorer regions (Molle, 2007). It is founded upon the conviction that such disparities threaten the integrity of the single market and are incompatible with the ideals of community and solidarity. The European Regional Development Fund was established in 1975. The Mediterranean enlargements (Greece in 1981 and Spain and Portugal in 1986) motivated the quest for regional and social cohesion. Importantly, the 1987 Single Market Act also included a section on economic and social cohesion and committed to reducing disparities between the various regions and to increase socio-economic levels of less-developed nations (Ireland, Greece, Spain and Portugal). In the space of the European Community of 12 Member States increasing regional differences had a north-south pattern, with Ireland in the western periphery. The Single Market Act recognised that excessive disparities between Member States and regions could be causing poorer Member States to block European legislation and impede implementation of various policies of the single market programme (Hix, 2005). One of the key principles of the cohesion and regional policy has become partnership between decision-makers in the core institutions of the EU, national governments and regional self-governments and administrations, together with representatives of labour unions, local business associations, and social action groups (Bachtler & McMaster, 2008). The cohesion and regional policies have allocated considerable funds and the principles of programming, implementation, monitoring and control have been characterised by increasing organisational complexity (Molle, 2007).

The costs of cohesion in the enlarging EU have been leading to questions about the future and considerable costs of regional policies in the EU15 and have been changing perceptions of this sector of policy-making (Baldwin & Wyplosz, 2006). Also the character of attitudes to environmental challenges has tended to change importantly since the beginning of the new millennium. Differences in socio-economic development between the Member States and regions and differences in environmental quality are associated with differences in national legislative responses to problems of cohesion and regional development and environmental problems. Such differences have been increased by the successive EU enlargements. In particular by the big May 2004 enlargement (with the Czech Republic, Cyprus, Estonia, Hungary, Latvia, Lithuania, Malta, Poland, Slovenia and Slovakia) and by the January 2007 enlargement (with Bulgaria and Romania) the regional and environmental disparities in the enlarged EU of 27 Member States increased significantly. The north-south gap in the EU tended to close, but the 2004 and 2007 enlargements have opened a new east–west gap, i.e. a gap in disparities between the 15 old Member States and 12 new Member States. It is therefore of little surprise that the increased disparities also tend to be reflected in significant differences in public opinion across the electorates of the EU27.

Questions are arising as to whether differences in public opinion represent important political cleavages across the enlarged EU with regard to future trends in environmental policies and the cohesion and regional policies of the EU27. Considering political divergence in public opinion across the countries concerned, it is also necessary to take into account the changing character of value orientations associated with the shift from an industrial society and its economic system towards a post-industrial society. The broad changes in value orientations result from life- style changes and have a significant bearing on perceptions of a relevant political agenda of the EU (Giddens, 2007). Therefore, it is crucial to understand that perceptions of environmental issues and issues of regional disparities articulated by citizens in current post-industrial societies tend to be different from the material survival concerns of industrial societies (Inglehart & Wenzel, 2005; Dostál, 2005). So-called post-materialist perceptions articulated in the public opinion in the post-industrial societies with their significantly modified environments tend to be based less upon direct experience of material survival, but much more upon abstract cognitive insights. The worldview is changing and is reflecting "a shift in what people want out of life" (Inglehart, 1997, p. 8). Moreover, the post-materialist value orientation also tends to be shaped by impacts of globalisation pressures on populations at local, regional and national levels and at the EU level. Such pressures result in new perceptions of the global system in terms of a "world risk society" and the EU is perceived as a "regional risk society" (Beck & Grande, 2007; Dostál, 2008).

Accordingly, the paper provides statistical analyses (based on correlation and principal component analyses, Rummel, 1970) of variations in public opinion on the EU environmental policies and regional and cohesion policies across the EU27. The analyses specify major divergence in public opinion and also indicate uncertainties and risks of an insufficient electoral support for the EU policy agenda in some Member States of the enlarged EU. The data analysed in this paper are derived from results of recent public opinion surveys (so-called Standard, Specific or Flash Eurobarometer surveys) organised by the European Commission. The main structure of the paper is as follows. First, using simplifying statistical procedures public opinion variations in view of globalisation and a post-materialist value orientation across the 27 polities of the enlarged EU are considered. Second, systematic variations across the EU27 in public opinion concerning trends of future environmental policies are specified. Third, variations in opinion on future trends of regional and cohesion policies are identified. Fourth, an explanatory correlation analysis in which variations on globalisation and in a post-materialist value orientation are used in a clarification of existing public opinion divergence across the enlarged EU concerning environmental policies and regional and cohesion policies is made. Finally, in the last section major conclusions from the analyses are drawn.

4.2 "Risk Society" and Public Opinion

It is clear that the changing perceptions of the EU political agenda must be seen in the context of a variety of globalisation pressures that stretch across the countries as a result of economic and social transformations of the current world system

(Held, McGrew, Goldblatt, & Perraton 1999). This has also been recognised in official documents of the European Commission such as the March 2000 Lisbon Agenda. The political elites of the EU15 committed themselves in Lisbon to a strategic agenda for the next decade of the EU to become the most dynamic and competitive knowledge-based economy in the world which would be able to avoid jobless growth and generate sustainable economic growth with more and better jobs, greater social cohesion and more respect for the environment. Obviously, significant differences in the perception and assessments of these various goals and relevant aspects of globalisation stretch from the EU and national political elites further to individual electorates of the enlarged EU. Therefore, Beck and Grande have warned that "the term 'jobless growth' remains trapped in the national outlook because it absolutises the national context and fails to realise that the creation of jobs is a transnational affair and must be analysed accordingly" (2007, p. 118). They have further argued that "the hierarchy of centre and periphery and the associated global inequalities are being inverted: the centre no longer represents the future and the prototype for the periphery" (2007, p. 119). Given the changing contexts of current perceptions of a relevant political agenda of the EU, it is therefore logical to assume that differences in view of globalisation across the 27 electorates of the enlarged EU can importantly contribute to explanation of differences in the attitudes to the future political agenda concerned with issues of environmental and cohesion and regional policies. Table 4.1 gives the results of an attempt to specify a dimension based upon opinions on five selected aspects of globalisation. Principal component analysis is applied (Rummel, 1970) which specifies basic dimensionality of the correlation matrix of five indicators. The indicators are answers from Eurobarometer no. 64 (fieldwork carried out in October and November 2005). The survey is based on samples of resident populations of the EU Member States aged 15 years and over. Basic sample design is a multi-stage random procedure and face-to-face interviews; the number of sampling points is drawn with probability proportional to population size (for a total coverage of the country) and to population density (NUTS II level). Sample size is 1,000 respondents in each country and in microstates Cyprus, Luxembourg and Malta 500 respondents.

Table 4.1 Dimension negative view of globalisation[1]

Indicators	Component loadings
(1) Currently afraid of job transfer to other Member States with lower production costs (QA18.8; mean = 68.5%)	0.909
(2) Globalisation leads to relocation of companies to countries where labour is cheaper (QA55; mean = 31.5%)	0.903
(3) Companies that relocate do so to increase profit (QA57; mean = 69.0%)	0.809
(4) Globalisation brings FDI to our country (QA56; mean = 15.4)	−0.602
(5) Net agreement that the EU protects us from negative effects of globalisation (QA56; mean = –7.5%)	−0.843

The results of the applied principal component analysis clearly indicate that positive loadings on the dimension represent opinions on the globalisation that emphasise anxiety about socio-economic impacts of globalisation. The highest loading on the component represents the opinion that people are currently afraid of job transfer to other Member States with lower production costs (component loading 0.909). It is significant to point out the very high mean level of this opinion in the EU27 (68.5%) indicating that this opinion is dominating. The next opinion explicitly says that globalisation leads to relocation of companies to countries where labour is cheaper (0.903). It is also significant to note that the mean is lower (31.5%) and this can convey an interesting message. It seems that electorates in some Member States do not believe in lower labour costs in countries which tend to gain in relocation. It is necessary to mention that this attitude tends to emphasise tensions in public opinion between the electorates in richer Member States with higher production costs and those in the new Member States with lower labour cost levels. A similar opinion tendency indicates the view that relocating companies do so to increase their profit (0.809). The very high mean value (69.0%) of the variable indicates that this opinion is also dominating. On the other side of the dimension, there is the negative loading of the net opinion recognising that the EU policies have capacities to protect citizens from negative effects of globalisation (–0.786). However, the mean value is negative (–7.5%) documenting that the share of negative answers to this question is larger than the share of positive answers. An optimistic view brings the belief that global economic relations enable inflows of foreign direct investment in the country concerned (–0.602). This view seems to express certain confidence in the country's competitiveness, but the mean value of this opinion is low (15.4%). It is obvious that this pattern of correlated views and their loadings on the specified dimension makes it possible to call the component a scale of *negative view of globalisation*. High scores of the EU countries on this dimension will represent anxiety and uncertainties concerning the globalisation pressures. Low scores will indicate opinion having more confidence in regard to current challenges of globalisation processes and their differentiating impacts in the enlarged EU. Obviously, this differentiation in the negative view of globalisation must be taken into account if the geo-economic context of attitudes to future environmental and cohesion policies has to be considered.

The same applies to differences across the enlarged EU in the shift towards post-materialist value orientations, because it can also be assumed that differences in post-materialist values can be considered as important public opinion factors having substantial effects on variations in public opinion about the future policy agenda concerned. It was already emphasised above that the shift toward post-materialism points to changing mass values and attitudes leading to decreasing importance of economic survival (materialism). It is associated with the structural shift from the era of industrialisation to the stage of post-industrial economy and society (Inglehart, 1997; Inglehart & Wenzel, 2005). This change implies increasing existential security in rich economies with advanced welfare-state provisions. It is important to emphasise in the context of this chapter that the shift towards post-materialist values and associated attitudes is resulting in life priorities of self-expression, and quality of life and, importantly, also in environmental concerns. Post-materialist value orientations also imply critical attitudes to authority, more critical and less easily led political opinion and a critical approach to the European integration processes (Dostál, 2002, 2006). It is therefore worthwhile to explore and specify the importance of differences in intensity of the post-materialist trend across the EU27. The individuals in the rich post-industrial societies who feel in material terms (i.e. economically) relatively safe and free to devote attention to concerns that are not immediately threatening them, tend to develop cognitive insights which produce risk awareness in more abstract terms of the "world risk society". In other words, it seems that with increasing economic safety there is associated decreasing egocentrism and increasing consciousness as regards environmental uncertainties and risks of increasing regional inequalities and these socio-cultural tendencies are reflected in changing public opinion in individual post-industrial countries of the global system and also across the enlarged European Union (see Dostál, 2008). The large number of surveys carried out in western post-industrial countries documented the shift from materialist survival value orientations towards post-materialist values that clearly appear to be more sensitive to environmental considerations. "Individual security increases empathy, making people aware of long-term risks. The rise of self-expression values fuels *humanistic* risk perception. These risk perceptions are fundamentally different from the *egocentric* threat perceptions that underline survival values" (Inglehart & Wenzel, 2005, p. 33). Accordingly, one can claim that the extent to which the post-materialist cultural expressions and perceptions of risk and environment and inequalities tend to prevail over materialistic survival values, reflects the levels of socio-economic development of countries concerned (see also Inglehart, 1997).

In Table 4.2 there are five indicators representing typical post-materialist and materialist opinions. These indicators are also derived from the survey of Standard Eurobarometer no. 64 carried out in the 27 countries. The structure of principal component loadings clearly shows the assumed distinction between post-materialist and materialist orientations. There are high positive loadings on the dimension of the emphasis given to protection of speech (0.935), demand of more information on environmental and nuclear safety policy (0.844) and to the priority of the EU to protect the environment (0.553). On the materialist side of the dimension there

Table 4.2 Dimension post-materialist value orientation[2]

Indicators	Component loadings
(1) Protecting freedom of speech (QA33a; mean = 10.1%)	0.935
(2) More informed on environmental and nuclear safety policy (QA22; mean = 26.1%)	0.844
(3) Priority of the EU to protecting the environment (QA34; mean = 22.2%)	0.553
(4) Priority of the EU to fighting unemployment (QA55; 43.7%)	−0.556
(5) Fighting rising prices (QA55; mean = 33.1%)	−0.835

are substantial negative loadings of materialist concerns with rising prices (–0.835) and the priority of the EU to fight unemployment (–0.556). It must be noted that the mean values of the materialist indicators are higher than the ones of the post-materialist opinion orientation.

The mean of 10.1% of the opinion on the importance of protecting freedom of speech is particularly low. These differences clearly document that, on the one hand, the shift towards the post-materialist value orientation is taking place but, on the other hand, this important cultural change is still in the current EU as a whole in an initial stage. However, in spite of this, some earlier public opinion research indicated that the shift towards the post-materialist value orientation is central to the understanding of variations in various other public opinion trends across the enlarged EU (Dostál, 2005, 2006). Hence, the component score on this dimension can be used to indicate differences in the post-materialist orientations across the 27 countries.

4.3 Opinion of EU Electorates on the Environment

The character of debates on environmental challenges has changed considerably since the beginning of the millennium (Antrop, 2008). The shift toward post-materialist values is bringing a change in the political agenda throughout the postindustrial and advanced industrial societies. The political agenda has been moved since 2000 away from a focus on economic growth at any price toward considerations of its environmental costs (Stern, 2007). This shift has brought a shift from major political divergence focused on socio-economic issues to political divergence based on cultural issues of life styles and quality of life concerns. In consequence, economic issues are increasingly sharing the political agenda of the Member States and the EU with issues which were less visible a generation ago. Giddens has argued that the policy area "where Europe could lead the world is the further development of ecological modernisation. It is possible that rather than further reducing competitiveness, the development of new ecological technologies – just as important, styles of life – could be a spur to its renewal" (2007, p. 187). He also has pointed out that important environmental issues are beset by risks that are still incalculable in terms of relevant actors and causal mechanisms and indeterminate in their societal and territorial impacts (Giddens, 2002).

Table 4.3 gives the results of a principal component analysis of seven chosen indicators. The indicators are derived from the survey of Special Eurobarometer no. 295 titled *Attitudes of European Citizens towards the Environment* and based on fieldwork carried out in November and December 2007. The analysis indicates that a major part of the correlations between the indicators (62.8%) can be represented by two orthogonal (i.e. not correlated and additive) components. The first component can be called a dimension representing concerns about *climate change* (34% of the total variance of the seven variables). The highest positive loading on

Table 4.3 Two components in public opinion on environment[3]

Indicators	Component 1 Climate change	Component 2 Landscapes/ disasters
(1) Talking about "the environment" one thinks first of pollution of towns and cities (Q2; mean = 18.9%)	−0.646	−0.308
(2) Talking about "the environment" one thinks first of climate change (Q2; mean = 16.3%)	0.904	−0.332
(3) Talking about "the environment" one thinks first of green pleasant landscapes (Q2; mean = 14.7%)	−0.274	0.781
(4) Worried about climate change (Q3; mean = 55.9%)	0.771	0.227
(5) Worried about water pollution – seas, rivers, underground water (Q3; means = 47.4%)	0.092	0.703
(6) Worried about air pollution (Q3; mean = 42.1%)	−0.691	−0.104
(7) Worried about natural disasters – earthquakes, floods, (Q3; mean = 32.9%)	0.005	0.705

the dimension has the answer that when one is talking about the environment one thinks first about climate change (loading 0.904). The second highest loading has the answer that respondents are worried about climate change (0.771). On the other pole of the dimension there are significant negative loadings of variables indicating the importance given to (air) pollution of towns and cities. This loading structure clearly documents a polarisation between, on the one hand, abstract considerations of climate change and, on the other, concrete concerns with pollution of air and local levels in towns and cities. It must be stressed that the mean level of worrying about climate change is relatively high (55.9%).

The second component can be called *landscapes and disasters* and this dimension is represented by almost 27% of the total variation of the correlation matrix of the used indicators. The highest positive loading on this component has the answer that when one is talking about the environment one thinks first about green and pleasant landscapes (loading 0.781). The second highest loading has the answer that respondents are worried about natural disasters – earthquakes, floods, etc. (0.705). A similar high loading has the answer that people are worried about water pollution – seas, rivers or underground water (0.703). The mean level of the association of environment with green pleasant landscapes is lower (14.7%), but the other two indicators represent higher average levels of environmental concerns (32.9 and 47.4%).

The outcomes of the principal component analysis document the significance of the two dimensions. The scores on the first indicate across the EU27 the more abstract concerns with climate change and global warming. Solving problems of climate change and global warming certainly demands EU-wide and especially world-wide collaboration. It seems that public opinion orientations considering climate change and global warming as crucial environmental concerns tend to perceive the current EU as a "regional risk society" that has to develop a political agenda that can be effective in the even wider context of the global system (Beck & Grande, 2007). In contrast, perceptions represented by the second component seem to be more concrete and contextual and locally and regionally constituted.

4.4 Opinion of EU Electorates on Orientations of Regional and Cohesion Policies

The Lisbon Agenda conveyed directly the issues of economic and social disparities between Member States and between regions. In the long period of the European integration process, the EC and EU underwent six successive enlargements in 1973–2007. The level of economic and social inequalities, both between Member States and regions, increased initially due to each enlargement. However, it was emphasised earlier in this paper that the 2004 and 2007 enlargements have substantially increased regional inequalities across the EU (Molle, 2007). The regional and structural funds and the cohesion funds are the key EU resources available to mitigate problems of disparities between Member States and regions. The funds contributed to economic and social development in regions and Member States concerned. The accession of the 12 new Member States in 2004–2007 did not result in increased budget contributions from the richer old Member States. Among other things, this can mean that reduction of disparities between regions and Member States can be difficult to achieve. It is therefore of little surprise that the European Commission also published Flash Eurobarometer no. 234 titled *Citizen's Perceptions of EU Regional Policy* based on fieldwork carried out in January 2008. The survey made an attempt to identify public opinion on orientations of regional and cohesion policies across the enlarged EU.

Table 4.4 gives the outcomes of another principal component analysis of ten selected indicators. The three rotated components represent together 70% of the total variation of the ten variables. The variables indicate what respondents consider as priorities important for their city or region. The respondents could choose ten priorities. The indicators are calculated as net positive opinions, i.e. the negative answers are subtracted from the positive answers.

The first component can be labelled as a dimension called *innovation* because it represents correlations between opinions giving priorities to EU regional policies orientated towards economic organisational innovation. The highest loading on the component has the priority of research and innovation (loading 0.847). The second highest loading has the priority given to support for small businesses (0.845). There follows a high loading of the priority of environment and risk prevention (0.730). There are further lower significant loadings representing the priorities given to energy infrastructure and sustainable energy supply and the priority of employment training (loadings 0.593 and 0.519). It is interesting to establish that the priority given to environment and risk prevention has the highest mean level of (77.3%) of these five variables.

The second component can be called *welfare* because it represents correlations between priorities given to maintenance of welfare-state provisions and the

Table 4.4 Three rotated components of public opinion on orientations of regional and cohesion policies[4]

Indicators	Component 1 Innovation	Component 2 Welfare	Component 3 Infrastructure
(1) Research and innovation (Q6C) mean = 38.4%	0.847	−0.252	−0.180
(2) Support for small businesses (Q6F) mean = 59.8%	0.845	0.371	0.016
(3) Environment and risk prevention (Q6E) mean = 77.3%	0.730	0.199	0.451
(4) Transport – rails, roads and airports (Q6A) mean = 53.6%	−0.176	0.042	0.925
(5) Information, communication technologies (Q6D) mean = 30.7%	0.268	0.264	0.601
(6) Energy, sustainable infrastructure (Q6B) mean = 48.8%	0.593	0.021	0.525
(7) Decisions EU projects in MS and regions (Q7) mean = 67.0%	0.069	0.745	0.044
(8) Employment training (Q6G) mean = 61.1%	0.519	0.698	0.154
(9) Involvement of local BA and TU (Q8) mean = 74.3%	−0.091	0.690	0.089
(10) Education, health and social infrastructure (Q6H); mean = 82.0%	0.156	0.670	0.542

importance of national and regional decision-making. The highest loading has a positive opinion on the right to decide about strategies and projects of EU regional policies in Member States and regions (loading 0.745). The second highest loading has the priority of employment training (0.698). A similar loading has also the positive opinion on the obligation of Member States and regions to involve local business associations and trade unions in considering equal opportunities and the institutional environment (0.690). The last significant loading on this component is the priority of education, health and social infrastructure (0.670). It is important to note that the four opinion variables also have high mean levels. Especially the mean of the priority of education, health and social infrastructure is very high (82%). This implies that this opinion orientation represents very significant perceptions of current regional and cohesion policies.

The third component is labelled *infrastructure* because it tends to represent opinion on priorities of various sorts of technical and social infrastructural policies. The highest loading on the dimension has the priority of better transport facilities ranging from railways and roads to airports (loading 0.925). The other loadings are lower: the priorities of information and communication technologies (0.601), education, health and social infrastructure (0.542), and energy infrastructure and sustainable energy supply (0.525). It is clear that this component also represents in an implicit way opinions on EU policy-making concerned with transportation, communication and social affairs (Molle, 2007).

4.5 Relationships Between Explanatory and Dependent Variables

Further explanatory analysis is based upon the correlation matrix which is shown in Table 4.5. In the preceding sections there have been suggested some possible factors and public opinion tendencies that can contribute to an explanation of the systematic differentiations in the above-specified perceptions of environmental issues and the priorities of EU regional policies. Table 4.5 provides the correlations (Pearson correlation coefficients) across the 27 Member States between two structural explanatory variables (GDP per capita in purchasing power standards in 2006 and the number of years of EU membership in 2007) and scores on the seven dimensions of public opinion specified with the help of the principal component analyses in the preceding sections. It has already been indicated that following earlier theoretical considerations the scores on the specified dimensions of the negative view of globalisation (see Table 4.1) and the post-materialist value orientation (see Table 4.2) are also used as explanatory variables.

The correlations given in Table 4.5 clearly demonstrate that some estimated relationships are substantial and also interesting in view of the earlier theoretical considerations. There are obviously zero correlations between the fifth and sixth dependent variables and between the seventh, eighth and ninth dependent variables because these two groups of component scores belong to the uncorrelated versions of the principle component analyses summarised in Tables 4.3 and 4.4. The relationships shown in Table 4.5 seem to allow the following interpretation.

Firstly, it is of little surprise that the GDP variable and the variable number of years of EU membership are significantly correlated (Pearson correlation coefficient

Table 4.5 Correlations between explanatory and dependent variables (Pearson correlation coefficients) (Sources: see Tables 4.1, 4.2, 4.3 and 4.4)

Indicators	(1)	(2)	(3)	(4)	(5)	(6)	(7)	(8)	(9)
(1) GDP per capita (in PPS) in 2006	1.00								
(2) Years of EU membership in 2007	0.68	1.00							
(3) PCA score negative view of globalisation	0.54	0.54	1.00						
(4) PCA score post-materialist values	0.55	0.43	0.50	1.00					
(5) PCA score worried about climate change	0.56	0.39	0.66	0.69	1.00				
(6) PCA score concerns on landscapes, etc.	−0.29	−0.62	−0.10	−0.38	0.00	1.00			
(7) PCA score regional policy – innovation	0.38	0.44	0.34	−0.01	0.37	0.05	1.00		
(8) PCA score regional policy – welfare state	−0.34	−0.26	−0.42	−0.65	−0.57	0.07	0.00	1.00	
(9) PCA score regional policy – infrastructure	−0.25	−0.41	−0.46	−0.28	−0.19	0.37	0.00	0.00	1.00

of 0.68), because the economies of older Member States still have higher levels of aggregate productivity per inhabitant (Baldwin & Wyplosz, 2006). However, there are substantial positive correlations (0.54 and 0.55) between the GDP variable and the scores on the dimensions of negative view of globalisation and postmaterialism. It is therefore clear that the concerns about globalisation pressures and the post-materialist value orientation are more intensive in the public opinion of the richer Member States. It is also interesting to establish that the globalisation measure is more closely related to the years of EU membership (0.54) than the post-materialism measure (0.43). This means that the anxiety and concerns about globalisation are more dominant in the perceptions of electorates in the older Member States. It is also important that the correlation (0.50) between the scores on the globalisation measure and the post-materialism measure is significant, but at a lower level than some other correlations of theoretical importance in Table 4.5.

Secondly, it has been postulated that the more abstract concerns with climate change and global warming (see Table 4.3) will be related to the post-materialist value orientation. This hypothesis is clearly sustained by the high correlation (0.69) between the two measures (see Fig. 4.1). There is also a significant correlation (0.66) of these concerns with the globalisation measure. The scatter diagram in Fig. 4.1 can suggest that there is an important relationship with the number of years of EU membership. However, such correlation (0.39) is much lower, particularly due to the outlaying positions of the opinion in Sweden (SE) and Denmark (DK). There is a higher correlation (0.56) with the GDP variable indicating that the concerns with climate change are more intensive in the public opinion of the richer Member States.

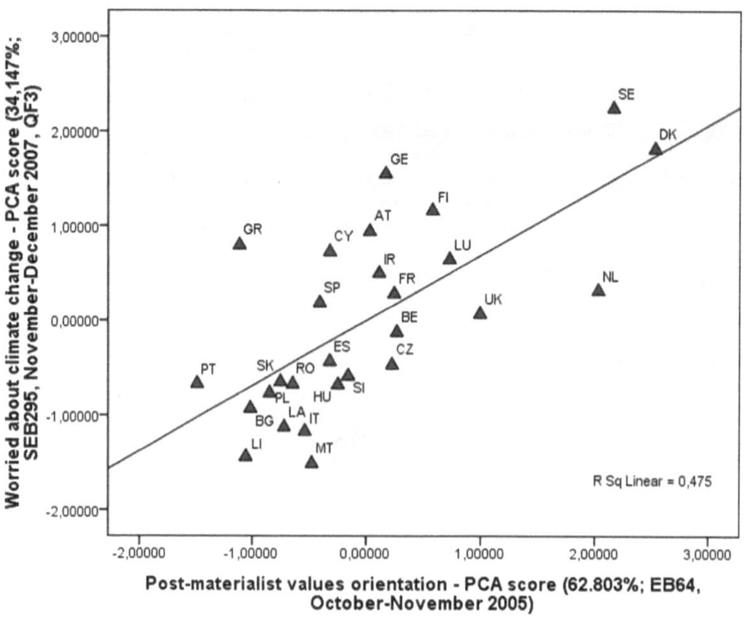

Fig. 4.1 Correlation between concerns about climate change and post-materialist scores

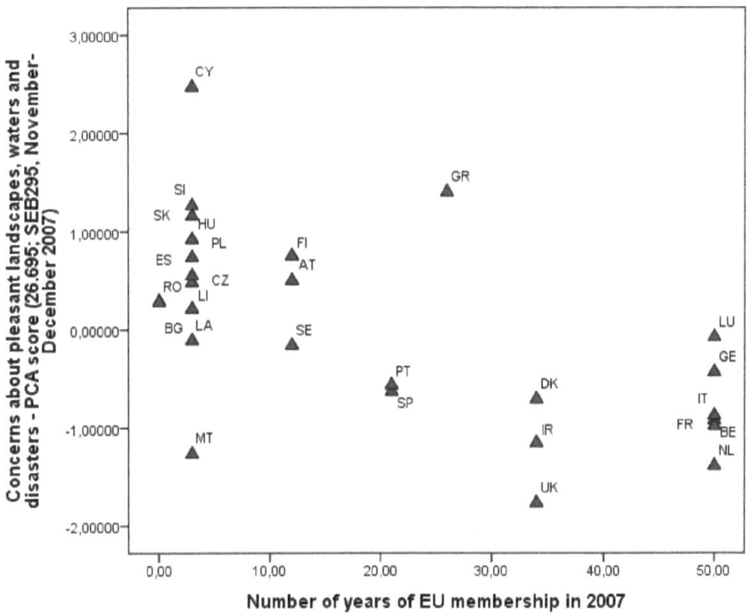

Fig. 4.2 Correlation between concerns about green pleasant landscapes and disasters and number of years of EU membership

Thirdly, there are in Table 4.5 also significant negative correlations. There is a strong negative correlation (–0.62) between the score on the landscape and disaster dimension and the number of years of EU membership. Figure 4.2 shows a more complex negative correlation documenting a clear dominance of the public opinion cleavage between, on the one hand, the opinion in the new Member States and, on the other hand, the opinion in the old Member States. The scatter diagram shows the extreme positions of Cyprus (CY) and Greece (GR) where some environmental circumstances (earthquakes, extensive woodland fires, etc.) are obviously shaping current perceptions of landscapes and natural disasters.

Finally, there are the correlations of the scores on the three components of opinions about priorities of regional and cohesion policies with the explanatory variables and the scores on the two environmental components. It appears that the opinion represented by the innovation component tends to be more intensive in richer countries (correlation of 0.38) and older Member States (correlation of 0.44). Interestingly, the opinion represented by the welfare component tends to show a number of negative correlations: with the GDP variable (–0.34), the globalisation measure (–0.42), the post-materialism measure (–0.65), and the climate change component (–0.57). These negative relationships suggest that this still very important opinion orientation (see the high mean levels in Table 4.4) on the maintenance of welfare-state provision and services and the role and obligations of national and regional authorities in projects of EU regional policies tends to be intensive in less-rich and newer

Members States with electorates that are less concerned about globalisation and follow largely materialistic value orientations.

The clear negative correlation with the scores on the dimension representing more abstract concerns with climate change and global warming is also in accordance with a dominance of materialistic perceptions which are not tending to associate environmental issues and region inequalities with the character of the "world risk society". Similar negative correlations characterise the relationships of the scores on the infrastructure component that specifies priorities given to technical and social infrastructural policies. However, the negative correlations are not high. There is an interesting low positive correlation (0.37) with the score on the landscape and disaster component indicating an association with environmental perceptions which characterise the public opinion in the new Member States.

4.6 Conclusions

It appears that the variations in public opinion represent important political divergence across the enlarged EU with regard to the future orientations of the environmental policies and the cohesion and regional policies of the EU27. It is also clear that the shift from an industrial society toward a post-industrial society has been resulting in current shifts in life styles and has important outcomes with regards to perceptions of a relevant political agenda of the EU. Therefore, it is crucial to understand that perceptions of environmental issues and issues of regional disparities articulated by citizens in current post-industrial societies tend to be different from the material survival concerns of industrial societies. The post-materialist perceptions articulated in the public opinion in the post-industrial societies tend to be based less upon direct experience of material survival, but much more upon abstract cognitive insights. The worldview is changing and reflects a change in what people want out of life. Moreover, the post-materialist value orientation also tends to be shaped by impacts of globalisation pressures on populations at local, regional and national levels and at the EU level. Such pressures result in new perceptions of the global system in terms of the "world risk society" and the EU is perceived as a "regional risk society". It seems that these tendencies are in part reflected by the public opinion orientations considering climate change and global warming as crucial environmental concerns perceiving the current EU as a "regional risk society" that has to develop a political agenda that can be effective in the even wider context of the global system. In contrast, the perceptions represented by the landscape and disasters component seem to be more concrete and contextual and locally and regionally constituted. The three components of public opinion on regional and cohesion policy trend's document a complex pattern that suggests more conservative value orientations and perceptions characterising more the industrial societies and less the post-industrial societies. Only the public opinion orientation on research and innovation, small businesses or environment and risk prevention tends to be a part of the emerging post-industrial era. The dominant trend is still to maintain within regional and cohesion policy the established welfare-state sevices and the role and obligations of national and regional authorities in regard to EU regional policies. This trend tends to be more dominant in less-rich and new Member States with electorates that are less concerned about globalisation and this indicates a largely materialist value orientation. A divergence of public opinion seems to emerge between the electorates of the old and richer historical core of the European Union and the electorates of the new Member States seemingly less aware of the global context in which the EU environmental and cohesion policies must take place considering global environmental problems and risks of global competition.

Acknowledgements Financial support by the Ministry of Education, Youth and Sport of the Czech Republic (MSM0021620831) is acknowledged.

Notes

1. N = EU27; source: Standard Eurobarometer 64, fieldwork: October–November 2005. European Commission, Brussels, June 2006; represented variance = 67.4%.
2. N = EU27; source: Standard Eurobarometer 64, fieldwork: October–November 2005. European Commission, Brussels, June 2006; represented variance = 63.6%.
3. N = 27; source: Special Eurobarometer 295, fieldwork: November–December 2007. European Commission, Directorate General Communication, Brussels, March 2008; represented variance by component 1 = 34.027% and by component 2 = 26.814%.
4. N = 27; source: Flash Eurobarometer 234, fieldwork: January 2008. European Commission, Directorate General Communication, Brussels, February 2008; rotation method is varimax with Kaiser normalisation; Total represented variance by the three components is 70%.

References

Antrop, M. (2008). Landscapes at risk: About change in the European landscapes. In P. Dostál (Ed.), *Evolution of geographical systems and risk processes in the global context* (pp. 15–33). Prague: P3K Publisher.

Bachtler, J., & McMaster, I. (2008). EU cohesion policy and the role of the regions: Investigating the influence of structural funds in the new member states. *Environment and Planning C: Government and Policy*, *26*(2), 398–427.

Baldwin, R., & Wyplosz, C. (2006). *The economics of European integration*. London: McGraw-Hill.

Beck, U., & Grande, E. (2007). *Cosmopolitan Europe*. Cambridge: Polity Press.

Dinan, D. (2005). *Ever closer union. An introduction to European integration*. Houndmills: Palgrave.

Dostál, P. (2002). EU enlargement and the public opinion on the Czech Republic: An explanatory analysis. *Geografie – Sborník České Geografické Společnosti*, *107*(2), 121–138.

Dostál, P. (2005). Uncertainties of public opinion on energy consumption across enlarged European Union: An explanatory analysis. *Acta Universitatis Carolinae – Geographica*, *40*(1), 25–45.

Dostál, P. (2006). Quo Vadis European Union? The core, peripheries and public opinion. *Acta Geographica Universitatis Comenianae*, *48*(1), 5–31.

Dostál, P. (2008). Changing geographical systems and risk processes: General considerations. In P. Dostál (Ed.), *Evolution of geographical systems and risk processes in the global context* (pp. 15–33). Prague: P3K Publisher.

Giddens, A. (2002). *Runaway world. How globalisation is reshaping our lives*. London: Profile Books.
Giddens, A. (2007). *Europe in the global age*. Cambridge, UK: Polity Press.
Held, D., McGrew, A., Goldblatt, D., & Perraton, J. (1999). *Global transformations. Politics, economics and culture*. Stanford, CA: Polity Press and Stanford University Press.
Hix, S. (2005). *The political system of the European Union*. Houndmills: Palgrave.
Inglehart, R. (1997). *Modernization and postmodernization. Cultural, economic, and political change in 43 societies*. Princeton, NJ: Princeton University Press.
Inglehart, R., & Wenzel, C. (2005). *Modernization, cultural change and democracy. The human development sequence*. Cambridge, UK: The Cambridge University Press.
McCornick, J. (2001). *Environmental policy in European Union*. Houndmills: Palgrave.
Molle, W. (2007). *European cohesion policy*. London: Routledge.
Rummel, R. J. (1970). *Applied factor analysis*. Evanston, IL: Northwerstern University Press.
Stern, N. (2007). *The economics of climate change. The stern review*. Cambridge: Cambridge University Press.

Chapter 5
Cross-Border Relationships of Small and Medium-Sized Businesses

Hartmut Kowalke, Olaf Schmidt, Katja Lohse, and Milan Jeřábek

5.1 Border Areas and Euroregions

Research on border areas is a traditional subject in geographical research (Bürkner, 1996; Arnold-Palussiére, 1983; Scott, 1999; Breysach, Paszek, & Tölle, 2003). After the political and economic changes in Middle and Eastern Europe in 1989/1990, themes concerning research altered completely (cf. Kowalke, Jerabek, & Schmidt 2004, 2005, 2008; Fassmann, 1997). The starting point was the functional change of borders and therewith, the border areas. Previously, there were practically closed borders separating Western Europe from Eastern Europe. Passenger traffic and exchange of goods were possible, most in the national, but not regional, interest until the borders opened for diverse exchange relationships. The 1st of May 2004 saw more changes in the quality of the borders. External borders converted into internal borders after the admission of Poland and the Czech Republic into the European Union. This had an impact on the border areas and the local businesses. On the 21st of December 2007, the Czech Republic joined the Schengen Agreement; thus systematic border controls of citizens were also abolished.

This new situation created advantages as well as disadvantages for the economy. On one hand, there were more opportunities for cross-border activities (new business relationships, bigger markets), however, on the other hand the opening of the borders caused problems (more rigorous competition, market domination).

Thanks to the trade-off and cooperation with the executive directors of the Euroregion Elbe/Labe in Pirna und Ústí nad Labem, the decision was made to prepare a questionnaire for small- and medium-sized businesses. 1.000 businesses were approached. Seventy-six German, and 52 Czech, companies agreed to participate in the survey which was then carried out with interviews. The main content of the survey was cross-border cooperation between businesses. The aim of the research report was to survey if and to what extent businesses of the Euroregion Elbe/Labe

H. Kowalke (✉)
Lehrstuhl für Wirtschafts- und Sozialgeographie Ost- und Südosteuropas, Technische Universität Dresden, 01062 Dresden, Germany
e-mail: hartmut.kowalke@mailbox.tu-dresden.de

J. Anděl et al. (eds.), *Landscape Modelling*, Urban and Landscape Perspectives 8, DOI 10.1007/978-90-481-3052-8_5, © Springer Science+Business Media B.V. 2010

cooperate with each other or if they plan to do so. Important were also questions concerning the awareness of possibilities for support, problems of collaboration which may appear, the valuation of the framework conditions, the implementation of the Euro and the role of the Euroregion (cf. Schamp, 1995; Grimm, 1996; Kowalke & Eckart, 1997; Bufon, 1998).

5.2 Method of the Survey

Our working group, which was composed of three scientists and 30 students, was planning to arrange 300 interviews with executives of small and medium-sized businesses in the area of Euroregion Elbe/Labe. In cooperation with the Chambers of Commerce and Industry of Dresden and Ústí nad Labem, approximately 1.000 businesses were chosen and contacted.

The data acquisition was carried out by questionnaire. Almost 500 German and 500 Czech small and medium-sized companies, located in the area of the Euroregion, were chosen. But only 76 German and 52 Czech companies agreed to participate in the survey.

The interviews were conducted with the help of students from the Geographical Institutes of the Technical University Dresden and the Purkyně-University Ústí nad Labem. The main questions were kept similar in order to obtain comparable results.

The questionnaire was structured in three thematically oriented complexes:

1. Complex: Framework conditions. The questions of this complex were targeted at the valuation of the conditions for cross-border cooperation (for example level of information about support programs).
2. Complex: Cross-border cooperation. In this complex, the aim was to distinguish between businesses which already cooperate, companies which are planning to build up a cross-border relationship and companies which are not planning to do so.
   - Already collaborating businesses were asked for their motives of building up such a relationship, the modalities concerning this cooperation and the problems that may appear including their perspectives when framework conditions change.
   - The aims of the questions for companies that are planning to cooperate were to find out their reasons, the status of preparation and the modalities of the planned cooperation.
   - The main interest regarding companies that don't wish to initiate cross-border cooperation is to find out their arguments for this.
3. Complex: characteristic of the business. The questions in this complex are related to each business, its size, its connection to an economic branch and its legal structure. They were also asked for an evaluation related to their business location and the Euroregion.

The preparation of the obtained data was carried out with the help of the program Microsoft Excel. The results are illustrated with charts and diagrams. Afterward some selected results were presented. They are concentrated on the second complex of the questionnaire.

5.3 Selected Results

The following overview (Table 5.1) demonstrates the composition of the participating businesses concerning their cross-border entrepreneurial activities.

Table 5.1 Participating businesses

	Czech businesses		German businesses	
	Number	Ratio in %	Number	Ratio in %
Businesses with existing cross-border cooperation	11	21.2	25	33.3
Businesses with a planned cross-border cooperation	11	21.2	21	28.0
Businesses which don't plan to build up a cross-border cooperation	30	57.7	29	38.7
Total	52	100.0	75	100.0

5.3.1 Businesses with an Existing Cross-Border Cooperation

As Table 5.2 shows the entrepreneurial relationships of German companies are more often based on contracts than those of the Czech companies. German businesses built their relationships from an earlier period.

Table 5.2 Types of business relations

	Czech businesses		German businesses	
	Number	Ratio in %	Number	Ratio in %
Informal, strategic cooperation without a contract	5	41.7	7	21.2
Contract-based cooperation for export/import	2	16.7	16	48.5
Sale and distribution per partner companies	1	8.3	3	9.1
Contracts based on the basis of subcontractors	1	8.3	3	9.1
Relocation of branch of production	2	16.7	3	9.1
Others	1	8.3	1	3.0
Total	12	100.0	33	100.0

Table 5.3 Sources of information about building cooperation

	Czech businesses		German businesses	
	Number	Ratio in %	Number	Ratio in %
Office of the Euroregion Elbe/Labe	0	0.0	3	6.8
Chambers of commerce and industry/chambers of skilled crafts and small businesses	2	14.3	3	6.8
Institutes for Economic Development	0	0.0	2	4.5
Public authorities	0	0.0		
Associations	0	0.0	2	4.5
Media (television, magazines etc.)	0	0.0		
Direct contacts via internet platforms	5	35.7	3	6.8
Fairs/events	0	0.0	7	15.9
Private contacts	6	42.9	15	34.1
Private consulting offices	0	0.0	2	4.5
Others	1	7.1	7	15.9
Total	14	100.0	44	100.0

[a]Multiple answers possible

German companies have more diverse sources of information available regarding the possibilities of building a cross-border cooperation than Czech companies (see Table 5.3). However, there are also similarities: the high usage of private contacts.

Most of the businesses have a positive view of the future relating to cross-border cooperation which is demonstrated in Table 5.4. One third of the German companies consider their cooperation as upgradable.

Out of all the possible motives supporting cooperation, the most important feature for businesses on both sides of the border is access to bigger markets. There are, however, mostly differences (see Table 5.5) as Czech business owners are mainly interested in exchanging information and gaining more experience. Any advantages in the possibility of reducing labour costs did not seem as important.

Table 5.4 Perspectives of the collaboration

	Czech businesses		German businesses	
	Number	Ratio in %	Number	Ratio in %
Positive	9	81.8	13	52.0
Negative	0	0.0	0	0.0
Unpredictable	1	9.1	4	16.0
Upgradable	1	9.1	8	32.0
Total	11	100.0	25	100.0

Table 5.5 Motives for the cooperation

	Czech businesses		German businesses	
	Number	Ratio in %	Number	Ratio in %
Tax benefits	0	0.0	2	4.3
Exchange of information and experiences	5	33.3	2	4.3
Access to new technologies	0	0.0	0	0.0
Access to a bigger market	4	26.7	10	21.7
Access to new resources	3	20.0	4	8.7
Diminishment of the costs of labour	1	6.7	6	13.0
Access to qualified manpower	0	0.0	1	2.2
Reduction of pressure due to concurrence	0	0.0	2	4.3
Use of transportation and logistic capacities	1	6.7	4	8.7
Increasing of degree of awareness	0	0.0	9	19.6
Tapping of subsidies	0	0.0	1	2.2
Others	1	6.7	5	10.9

[a]Multiple answers possible

Most of the business owners, especially from the Czech Republic expected a positive impact after the abolition of current restrictions on workers and service movements. But the answers show uncertainty concerning this abolition. Approximately every second German business owner is not able to estimate the effects of this abolition (Table 5.6).

Most of the polled companies recommend the fastest possible abolition of the restrictions. As Table 5.7 shows one quarter of the business owners want to keep

Table 5.6 Transition period for workers and service movements

	Czech businesses		German businesses	
Effect of the abolition of restrictions	Number	Ratio in %	Number	Ratio in %
Positive	5	45.5	7	29.2
Partly positive/partly negative	0	0.0	6	25.0
Negative	0	0.0	1	4.2
Unpredictable	6	54.5	10	41.7
Total	11	100.0	24	100.0

Table 5.7 Date of expiration of the restrictions of workers and service movement

	Czech businesses		German businesses	
	Number	Ratio in %	Number	Ratio in %
In 1–2 years	4	36.4	9	47.4
In 2–4 years	4	36.4	4	21.1
In 4–6 years	0	0.0	1	5.3
In 7 years	3	27.3	4	21.1
Equal	0	0.0	1	5.3
Total	11	100.0	19	100.0

Table 5.8 Expected impact of the implementation of the Euro in the Czech Republic

	Czech businesses		German businesses	
Impacts	Number	Ratio in %	Number	Ratio in %
Positive	0	0.0	...	56.0
Partly positive/partly negative	6	54.5	...	4.0
Negative	3	27.3	...	12.0
Unpredictable	2	18.2	...	28.0
Total	11	100.0	...	100.0

up the restrictions in the mid-term. These included mainly service businesses which expect negative effects due to falling prices or pressure from competitors.

Regarding the expected effects of the implementation of the Euro in the Czech Republic (mentioned in Table 5.8), the opinions of business owners differ immensely. Most German companies have a positive outlook. Czech companies are more sceptical, possibly because they anticipate a price increase.

5.3.2 *Businesses Planning Cross-Border Cooperation*

The reasons for hesitating to build cross-border relationships between companies are very diverse. Nonetheless, Table 5.9 shows that a lack of available information

Table 5.9 Reasons for businesses hesitating to support cross-border cooperation

	Czech businesses		German businesses	
	Number	Ratio in %	Number	Ratio in %
Lack of information about appropriate branches	3	17.6	5	8.8
Lack of information about appropriate partner companies	5	29.4	10	17.5
Lack of information about possible support	1	5.9	5	8.8
Currently bad economic conditions	0	0.0	2	3.5
Effort higher than benefit	1	5.9	8	14.0
Lack of appropriate partner companies	3	17.6	8	14.0
Inadequate infrastructure for cross-border traffic	0	0.0	2	3.5
Negative experiences	0	0.0	2	3.5
Legal barriers	0	0.0	1	1.8
Trade barriers (p. e. customs duty, border formalities)	1	5.9	1	1.8
Language barriers	1	5.9	8	14.0
Negative impact on image	0	0.0	1	1.8

[a]Multiple answers possible

Table 5.10 Sources of information for the planned cooperation

	Czech businesses		German businesses	
	Number	Ratio in %	Number	Ratio in %
Office of the Euroregion Elbe/Labe	1	6.7	4	8.3
Chambers of commerce and industry/chambers of skilled crafts and small businesses	5	33.3	6	12.5
Institutes for Economic Development	0	0.0	6	12.5
Public authorities	0	0.0	2	4.2
Associations	0	0.0	3	6.3
Media (television, magazines etc.)	0	0.0	2	4.2
Direct contacts via internet platforms	4	26.7	4	8.3
Fairs/events	2	13.3	13	27.1
Private contacts	3	20.0	5	10.4
Private Consulting offices	0.0	0.0	2	4.2
Other	0	0.0	1	2.1
Total	15	100.0	48	100.0

[a]Multiple answers possible

has had a clear impact, as has a lack of appropriate partner companies. German companies expect language will be a barrier and are wary that they may have to put in a lot of effort for very little benefit.

A wide variety of sources of information are available to German companies (compared to companies with existing relationships). Czech companies have limited resources regarding accessing information (see Table 5.10). More preparation time for German companies could be crucial in this case.

5.3.3 Businesses Refusing to Build Cross-Border Cooperation

Table 5.11 is interesting because two thirds of the German companies polled refuse to collaborate across borders but do support the idea. Czech companies are very

Table 5.11 Willingness to cooperate cross-border

	Czech businesses		German businesses	
Impacts	Number	Ratio in %	Number	Ratio in %
Yes	6	20.0	20	66.7
No	1	3.3	1	3.3
Not applicable for me	9	30.0	5	16.7
I don't know	14	46.7	4	13.3
Total	30	100.0	30	100.0

Table 5.12 Reasons why cross-border cooperation is not planned

	Czech businesses		German businesses	
	Number	Ratio in %	Number	Ratio in %
Lack of information about appropriate branches	4	7.8	3	3.8
Lack of information about appropriate partner companies	7	13.7	7	8.9
Lack of information about possible support	3	5.9	5	6.3
Currently bad economic conditions	2	3.9	5	6.3
Effort higher than benefit	7	13.7	12	15.2
Lack of appropriate partner company	7	13.7	4	5.1
Inadequate infrastructure for cross-border traffic	0	0.0	1	1.3
Negative experiences	0	0.0	5	6.3
Barriers by law	2	3.9	6	7.6
Trade barriers (p. e. customs duty, border formalities)	1	2.0	3	3.8
Language barrier	5	9.8	9	11.4
Fear of damaged image	0	0.0	1	1.3
No interest (as of yet)	8	15.7	11	13.9
Other	5	9.8	7	8.9
Total	51	100.0	79	100.0

[a]Multiple answers possible

uncertain about the concept. For some businesses there are no available contacts for this kind of collaboration.

In Table 5.12 a variety of reasons for refusing cross-border cooperation include: objective reasons (e.g. lack of information, language barriers) as well as subjective reasons (lack of interest). It is therefore impossible to restrict intensifying entrepreneurial relationships to one or two different methods. In this case, different approaches are needed.

5.4 Conclusions

The aim of the research report was to survey if, and to what extent, businesses from the Euroregion Elbe/Labe are prepared to cooperate with each other and to analyse which problems of collaboration may appear and which circumstances may hamper the cooperation (cf. Krätke, 1998).

An increase in adequate information about possible collaboration and the support of an appropriate cooperation partner are required for creating successful relationships (Kowalke, Jerabek, Schmidt, & Lohse, 2008; Jeřábek, 2002). Despite a high amount of available information, most of the polled companies felt they needed more. Considering the current increase in interest about building cooperation, additional support and available information would be of great benefit. This would include more contributions from the relevant offices within the Euroregion (cf. EUREK, 1999).

Financial support from different programs, such as the Structural and Community Fund of the European Union, has been low due to a lack of information supplied to them. Most companies consider the offer of support could provide encouragement but this would not be seen as a reason to build entrepreneurial relationships. In fact, such cross-border cooperation provides business owners with possible competitive advantages due to the extension of their market area and a greater degree of awareness.

Saxon companies anticipate a cost benefit as a result of the cross-border cooperation due to lower labour costs (including non-wage labour costs). In contrast, the most important reasons for Bohemian companies to build ties are the exchange of experience and information. Other obstacles for fostering cooperation are missing information about potential partner companies, and the language barrier.

The results, including the entrepreneurial perspectives, are evaluated by the polled companies as "very good" or "good". The fact that a large proportion of businesses have already developed international contacts, demonstrates that companies consider cross-border relationships essential for their economic success.

For the business owners, inviting the Czech Republic to join the European Union could provide advantages such as; the abolishment of obligatory customs declarations, as well as simplifying collaboration and communication. The Schengen Conventions added more areas and is a further step towards the intensification of cross-border cooperation of companies.

Acknowledgements This publication is a further step into the long-lasting and fruitful cooperation of colleges and students of the Geographical Institutes of the universities of Dresden and Ústí nad Labem. We would also like to thank the executive directors of the Euroregion Elbe/Labe in Pirna und Ústí nad Labem, because without cooperation this research project would not have been possible.

References

Arnold-Palussiére, M. (1983). *Die grenzüberschreitende regionale zusammenarbeit auf dem gebiet der raumordnung*. Hannover: ARL.

Breysach, B., Paszek, A., & Tölle, A. (Eds.). (2003). *Grenze – Granica/Interdisziplinäre Betrachtungen zu Barrieren, Kontinuitäten und Gedankenhorizonten aus deutsch-polnischer Perspektive. Wissenschaftliche Reihe des Collegium Polonicum, Bd. 8*. Berlin: Europa-Universität Viadrina Frankfurt (Oder), Uniwesytet im. Adama Mickiewicza w Poznaniu, Logos Verlag.

Bufon, M. (1998). Border and border landscapes: A theoretical assessment. In M. Koter & K. Heffner (Eds.), *Region and regionalism No. 3: Borderlands or Transborder regions – Geographical, social and political problems* (pp. 7–14). Lodz: University of Lodz and Silesian Institut in Opole.

Bürkner, H. J. (1996). Geographische grenzraumforschung vor neuen herausforderungen – forschungskonzeptionen vor und nach der politischen wende in Ostmitteleuropa. In

H. J. Bürkner & H. Kowalke (Eds.), *Geographische Grenzraumforschung im Wandel* (Vol. 15, pp. 1–12). Potsdam: Praxis Kultur und Sozialgeographie.

EUREK. (1999). *Europäisches Raumentwicklungskonzept (European Spatial Development Perspektive – ESDP). Auf dem Wege zu einer räumlich ausgewogegenen und nachhaltigen Entwicklung der EU*. Potsdam.

Fassmann, H. (Ed.). (1997). *Die rückkehr der Regionen. Beiträge zur Regionalen Transformation Ostmitteleuropas*. Wien: Verlag der Österreichischen Akademie der Wissenschaften.

Grimm, F. D. (1996). Diskrepanzen und verbundenheiten zwischen den deutschen, polnischen und tschechischen grenzregionen an der lausitzer neiße ("euroregion neiße"). *Europa Regional, 4*(1), 1–14.

Jeřábek, M. (2002). Crossborder cooperation and development in Czech borderland. *Acta Universitatis Carolinae. Geographica, 37*(1), 45–60.

Kowalke, H. & Eckart, K. (Eds.). (1997). *Die Euroregionen im Osten Deutschlands. Schriftenreihe der Gesellschaft für Deutschlandsforschung 35*. Berlin: Duncker – Humblot.

Kowalke, H., Jerabek, M., & Schmidt, O. (Eds.). (2004). *Grenzen öffnen sich – Chancen und Risiken aus Sicht der Bewohner der Sächsisch-Böhmischen Grenzregion. Dresdner Geographische Beiträge* 10. Ústí nad Labem: Technische Universität Dresden, Dresden and UJEP.

Kowalke, H., Jerabek, M., Schmidt, O., & Lohse, K. (Eds.). (2008). *Grenzüberschreitende Beziehungen von klein- und mittelständischen Unternehmen der Euroregion Elbe/Labe. Dresdner Geographische Beiträge 12*. Dresden: Technische Universität Dresden.

Kowalke, H., Schmidt, O., & Jerabek, M. (2005). Entwicklungsprozesse und Entwicklungsprobleme im sächsisch-böhmischen Grenzraum. In E. Mehnert (Ed.). *. . . 's kommt alles vom Bergwerk her, Materialienband zum 7*, Deutsch-Tschechischen Begegnungsseminar, Frankfurt/Main, Berlin, pp. 147–159.

Krätke, S. (1998). Problems of cross-border regional integration: the case of the German-Polish border region. *European Urban and Regional Studies, 5*(3), 249–262.

Schamp, E. W. (1995). Die bildung neuer grenzüberschreitender regionen im östlichen mitteleuropa – eine Einführung. *Frankfurter Wirtschafts- und Sozialgeographische Schriften, 67*(1), 1–18.

Scott, J. W. (1999). European and North American contexts for cross-border regionalism. *Regional Studies, 33*(7), 605–617.

Chapter 6
Land-Use Changes Along the Iron Curtain in Czechia

Ivan Bičík, Jan Kabrda, and Jiří Najman

6.1 Changing Functions of Czech Borders

Changes of land-use structure are influenced by a wide range of factors, the so-called "driving forces" (Jeleček, 2002). According to Mather (2002) or Lambin and Geist (2007) these include economy, technology, politics, institutions and culture work on a general level – as "underlying" factors. On the contrary, there exist "proximate" factors, working on a local level. We have basically two main clusters of proximate factors – natural conditions (altitude, slope, soil fertility, climate, etc.) and socio-economic characteristics (e.g., density of population, economic structure, and spatial exposedness). Another proximate factor, influencing local land use, is proximity to or position relative to political borders.

The borders in Czechia played very different roles in the last century. At the beginning, in the period of the Austro-Hungarian monarchy, the role of borders was relatively weak, influenced by the fact that the Czech lands were a part of a much larger political unit, in which the same economic and custom laws played roles. After Czechoslovakia was established in 1918, the same borders started playing different roles both in economy and in politics. In the years 1939–1945, old borders ceased to exist, and the Nazi Protectorate Bohemia and Moravia (Böhmen und Mähren) was established. A short period of democracy controlled by communists (1945–1948) followed. From February 1948 onwards, the Iron Curtain was erected step-by-step on the western borders of Czechoslovakia. This Iron Curtain (ca. 1948–1990) between the European East and West determined land-use changes in those Czech regions lying along the borders with Austria and the former West Germany.

The Iron Curtain was not only an abstract line, named so by Sir Winston Churchill, but also a very concrete barrier with fences, walls and guard posts. Large areas along it had to be abandoned, depopulated, afforested, and they served as military zones where the Czechoslovak army "faced" the phantom enemy from the "evil

I. Bičík (✉)
Department of Social Geography and Regional Development, Charles University in Prague, Albertov 6, 128 43 Praha 2, Czech Republic
e-mail: bicik@natur.cuni.cz

J. Anděl et al. (eds.), *Landscape Modelling*, Urban and Landscape Perspectives 8, DOI 10.1007/978-90-481-3052-8_6, © Springer Science+Business Media B.V. 2010

capitalist West". A special zone was the area between the Iron Curtain and the border itself, which was totally inaccessible and excluded from any economic use (e.g. Bičík & Štěpánek, 1994). Here, the afforestation was mostly spontaneous.

The end of the year 1989 created conditions for an abolishment of strongly controlled borders with both "Eastern" and "Western" neighbours of Czechia (Germany – East and West earlier, Austria, Poland, and most recently Slovakia from 1993). In the period 1990–2004, many new checkpoints were installed on Czech borders, especially on highways and motorways. Finally, borders lost the old functions they played during the last century, and gained a new role, especially after Czechia joined the EU and the Schengen Treaty.

All these factors should result in a specific land-use structure and its changes in border regions. Border regions should be used less intensively, i.e. they should be exposed to a weaker anthropogenic impact – with less arable land and areas connected with urbanisation (built up areas, gardens, etc.), and more forested areas, permanent grasslands and abandoned or unused land. Furthermore, border regions should exhibit a stronger extensification (decrease of anthropogenic activity), especially in time periods of radical political influences, for example when the borders were closed by the Iron Curtain (Štěpánek, 2002).

During the period of one century, the borders of the new country created in 1918 changed their functions many times. This fact led us to forming several research hypotheses. How did the border line with different neighbours of Czechia influence the landscape in its surroundings in different time periods? How was land-use structure influenced by borders in different conditions in comparison with the interior of Czechia? How did the economic and political transformation after 1990 influence land-use structure in border regions?

In the past, we realized several studies of border regions using statistical data on land-use changes (Bičík & Štěpánek, 1994; Štěpánek, 2002; Bičík & Kabrda, 2008). In this article, we examine land-use changes in border regions of Czechia during the ten years of transformation (1990–2000) in comparison with the older period 1948–1990 (centrally planned economy). For better understanding of processes occurring in the Czech border landscape after the fall of the Iron Curtain, we will compare the results obtained for the years 1990 and 2000 with land-cover data from remote sensing – from the CORINE land-cover database (using LANDSAT images) for 1990 and 2000.

6.2 Data Sources and Methods

Our research is based on two different data sources: cadastral statistics (LUCC UK database), and the CORINE land-cover database – see the following text in this chapter. Accordingly, methods, and studied territories and time horizons differ as well.

The first data source is the *database of long-term land-use changes in Czechia* (LUCC UK Prague, http://lucc.ic.cz). It is derived from cadastral statistics (see

Bičík, Jeleček, & Štěpánek, 2001 or Bičík & Jeleček, 2003 for more details). In this database, the whole area of Czechia is divided into 8,903 Basic Territorial Units (BTUs), each consisting of one or more cadastres. Land-use structure of each BTU is recorded in four time horizons, representing the main milestones of modern Czech history – 1845 (before deeper impacts of market economy on land-use structure), 1948 (communist coup), 1990 (the "Velvet Revolution") and 2000 (after ten years of transformation). Eight basic land-use categories are recognised in this database: arable land, permanent cultures (gardens + vineyards + orchards + hop gardens), meadows and pastures (together permanent grasslands), forested areas, water areas, built up areas and remaining areas (non-productive land, bare land, infrastructure, mines, etc.).

For the purpose of our research, we have defined several subsets of BTUs in relation to their proximity to the state borders (see Bičík & Kabrda, 2008 for more details). Firstly, we have defined three "belts" of BTUs along the borders of Czechia. Subset "At border" (A) consists of BTUs adjoining/touching the border ($n = 395$). Subset "Intermediate" (B) consists of BTUs adjoining the BTUs of subset A ($n = 571$). Subset "At interior" (C) consists of BTUs adjoining the BTUs of subset B, excluding the BTUs of subset A ($n = 580$). The remaining BTUs create subset "Interior" (O, $n = 7,357$). The first three subsets (belts of BTUs) were then merged (A + B + C, $n = 1,546$) and divided according to respective countries. Five subsets were defined in this way – former West Germany ($n = 288$), former East Germany ($n = 316$), Poland ($n = 491$), Slovakia ($n = 168$) and Austria ($n = 283$). In this article, only two subsets with the countries divided from Czechia by the Iron Curtain (West Germany and Austria) were studied, and compared to the characteristics of the Interior (see Fig. 6.1 for graphical expression).

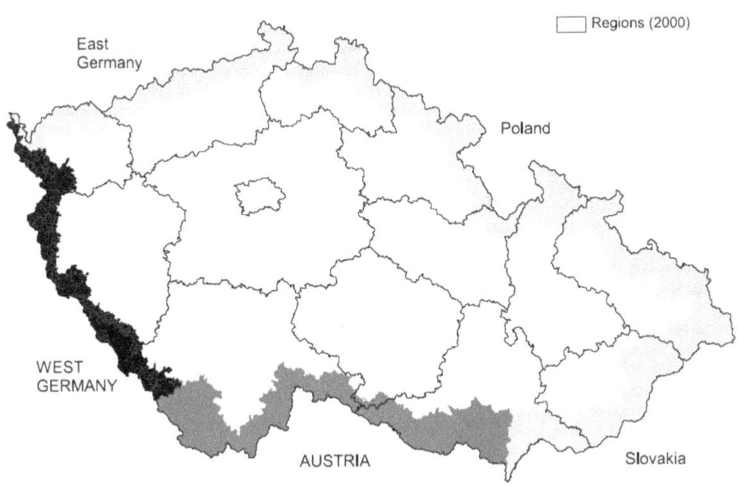

Fig. 6.1 Delimitation of subsets of BTUs according to their position towards borders (Source: own calculations)[1]

Table 6.1 Basic characteristics of the subsets West Germany, Austria and Interior (Source: LUCC UK Database, own calculations)[2]

	West Germany	Austria	Interior
Average price of agricultural land (1992, CZK/m^2)	2.14	4.23	4.34
Average altitude (metres above sea level)	704.60	479.70	415.60
Average slope (°)	4.25	2.41	2.48
Share of area (%) in the Sudetenland	87.50	68.50	22.50
Density of population (1991, inhabitants/km^2)	49.40	55.60	136.60
Share of area (%) of BTUs in peripheral spatial position	82.80	83.90	20.40

Land-use structure was calculated for these subsets for the years 1948, 1990 and 2000 (Figs. 6.4, 6.5 and 6.6). For better understanding of land use and its changes in different subsets, several "proximate" characteristics influencing land use were calculated (Table 6.1), both natural (average altitude and slope, official price of agricultural land as a complex indicator of suitability of a territory for farming – e.g. Kabrda, 2004) and socio-economic (density of population, share of area influenced by the expulsion of Czech Germans after WWII, and share of area in peripheral spatial position according to Hampl, Gardavský, & Kühnl, 1987). All the data were extracted from the LUCC UK database.

The second data source used for this study is based on the LANDSAT images from the years 1990 and 2000 – see Najman (2008) for more information. These data were used in the form of the *CORINE Land Cover (CLC) database* from EEA (European Environment Agency). We used the project IMAGE2000 and its three levels: Corine Land Cover 1990, Corine Land Cover 2000 and Corine Land-Cover Changes. Minimal pixel in the first and second sources is 25 ha, but only 5 ha in the third one. From the CORINE database, we used 14 categories of land cover, which were then merged into seven categories (arable land, permanent cultures, grasslands, forest land, built up areas, water areas, other areas). This simplification gives us a better chance to compare changes obtained from remote sensing with the above-described data from cadastral statistics. Data from CLC were evaluated in ESRI ArcGIS 9.2 software. Entered data were reclassified using the extension programme Spatial Analyst and the function Zonal Statistics.

Thus, this methodology is based on a comparison of two satellite images (or their interpretation in Corine, respectively) from the years 1990 and 2000. We studied land cover and its changes in a belt on both sides of the former Iron Curtain, 15 km wide to each side (hence 30 km in total), stretching from the Baltic Sea to the borders of former Yugoslavia. This belt was divided according to respective countries, and further on its western (WEST) and eastern (EAST) part, in order to compare land cover on both sides of the former Iron Curtain.

With the help of these two images, and with field mapping, we also studied land cover in selected National parks – one on the Czech/German border (Šumava and Bayerisches Wald), and another on the Austrian/Hungarian border (on the shores of Neusidler see).

6.3 Land-Use Changes Along the Iron Curtain in Czechia on the Level of BTUs (1948–1990–2000)

In the following, we will describe and explain land-use changes between the years 1948, 1990 and 2000 in the regions of Czechia bordering the former West Germany and Austria, and compare them with the interior of the country (Fig. 6.1). Obviously, land use in border regions was influenced not only by their proximity to the Iron Curtain, but by natural and socio-economic conditions as well. Thus, Table 6.1 contains selected indicators for subsets Austria and West Germany in comparison with the interior of the country.

Table 6.1 proves that regions bordering West Germany, when compared to the interior, are typical of worse natural conditions (lower official price of agricultural land, higher average altitude and slope) and weaker socio-economic activity (lower density of population, higher share of BTUs with peripheral position, almost 90% of area affected by the expulsion of Germans). The situation in regions bordering Austria is slightly different. Their socio-economic characteristics are similar to the subset West Germany – although the share of BTUs influenced by the expulsion of Czech Germans after World War Two is lower – but their natural conditions seem to represent the average within Czechia.

However, Table 6.1 presents only mean values for the whole subsets of BTUs. If we examine the internal heterogeneity of these characteristics in the studied subsets, we reveal significant differences.

The region bordering former West Germany is highly homogenous – almost totally constituted of mountains and highlands (e.g., Český Les, Šumava or Smrčiny) and suffering from strong depopulation during the whole Twentieth century (except some minor localities like Chodsko), especially after WWII – see the maps and text in e.g. Štěpánek (2002) or Bičík and Kupková (2002).

On the contrary, the BTUs bordering Austria are highly heterogeneous both in their natural conditions (Fig. 6.2) and socio-economic activity (Fig. 6.3). This region can be roughly divided into two parts (Chromý & Rašín, 2009). The eastern part lies in the lowlands along the Dyje river, with fertile soils and favourable climate. Its landscape is intensively used for productive agriculture. On the other hand, the western part of subset Austria reminds us of the regions bordering West Germany. It is covered with highlands (e.g., Novohradské hory and Českomoravská vrchovina) and mountains (Šumava) with less favourable natural conditions. Consequently, the density of population is much lower, as well as the general anthropogenic pressure on the landscape. We cannot omit this sharp dichotomy of the subset Austria

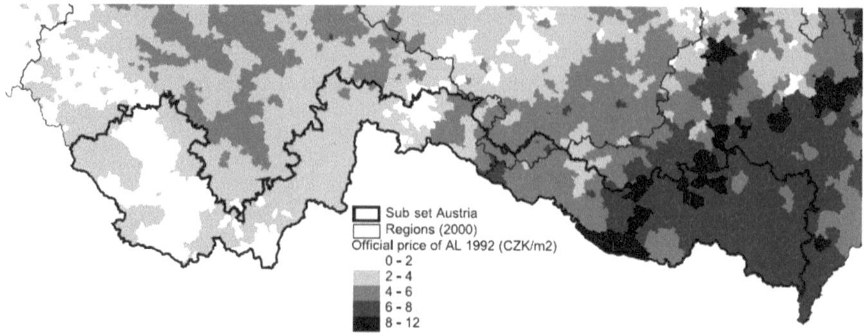

Fig. 6.2 Official price of agricultural land (1992, CZK/m^2) in the BTUs of subset Austria (Source: CORINE Land-cover database, own calculations, Najman, 2008)

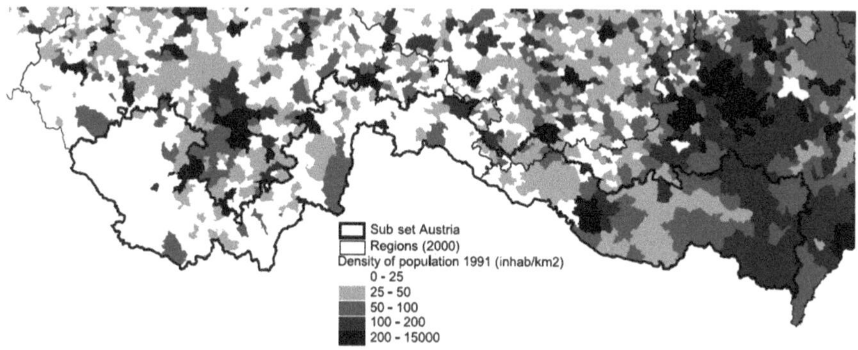

Fig. 6.3 Density of population (1991, inhabitants/km^2) in the BTUs of subset Austria (Source: CORINE Land-cover database, own calculations, Najman, 2008)

when interpreting land use, for it smoothes and averages differences between the intensively used landscape in its eastern part and relatively extensively used one in the western part.

Land-use changes in the subsets of BTUs bordering West Germany (Fig. 6.4) and Austria (Fig. 6.5) were studied in two time periods – socialistic (1948–1990) and restoration of market forces (1990–2000) – and compared to those in the interior of the country (Fig. 6.6). Several conclusions can be draw from these figures.

Land use generally correlates with both natural and socio-economic conditions in these regions. The landscape of BTUs bordering West Germany is used very extensively, with an extreme share of forests (over 55% in 2000), above-average share of permanent grasslands (over 17% in 2000) and a negligible share of arable land (15% in 2000). Land use in the subset Austria has greater similarity to that of the interior of the country (compare Figs. 6.5 and 6.6), although still less intensive (share of arable land lower by 7 pp and share of forested areas higher by 6 pp in 2000). It is also typical of an above-average share of permanent cultures – especially vineyards in its eastern part (e.g. Pálava hills); and water areas (4% of the whole

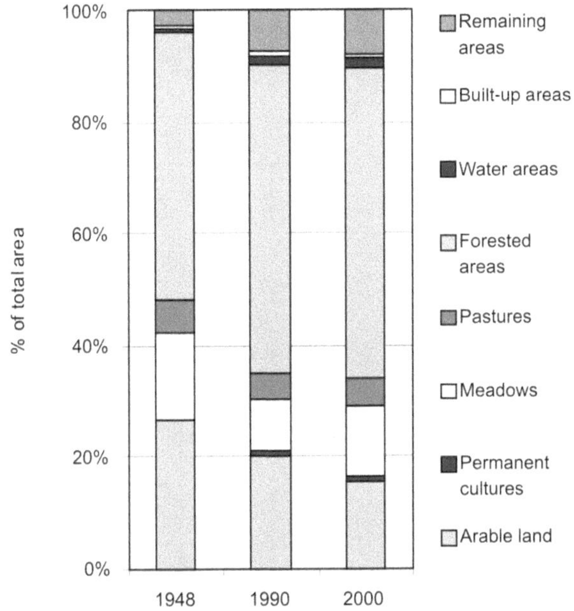

Fig. 6.4 Land-use changes in the subset West Germany ($n = 288$) (Source: CORINE Land-cover database, own calculations, Najman, 2008)

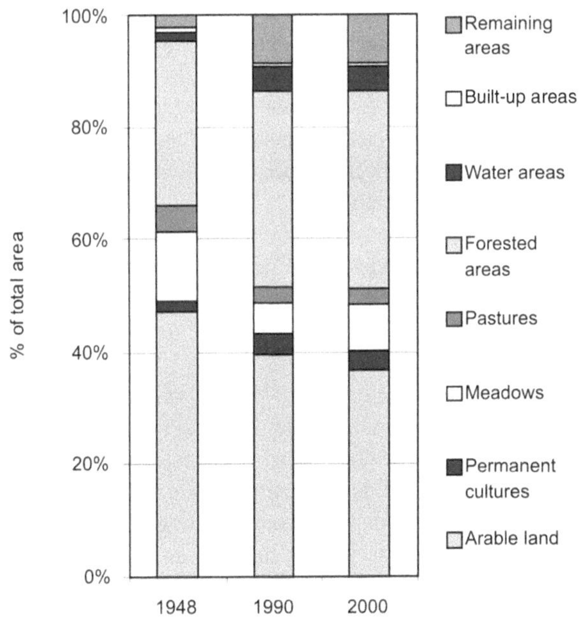

Fig. 6.5 Land-use changes in the subset Austria ($n = 283$) (Source: LUCC UK Database)

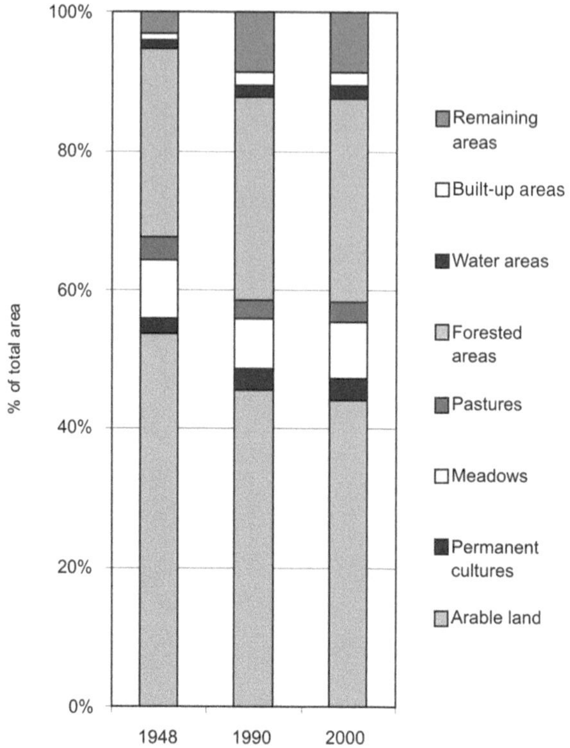

Fig. 6.6 Land-use changes in the subset Interior ($n = 7{,}357$) (Source: LUCC UK Database)

area in 2000), consisting both of fishing ponds and large reservoirs (e.g., Lipno to the west and Nové Mlýny to the east). The share of built up areas is below average in both subsets, demonstrating lower socio-economic activity and absence of larger cities and concentrations of population.

Land-use changes had the same direction in both border subsets as in the interior of the country in both time periods, implying a general character of these trends – compare to Jeleček (2002) for the whole of Czechia; see ibid for explanation and "driving forces" of these changes. However, the intensity of these changes was different.

The first – socialist – time period (1948–1990) was typical of decreasing share of arable land and permanent grasslands, and increasing share of forested, built up and "remaining" areas. On the one hand, these processes were the result of growing intensity of agriculture (with yields on fertile plots growing faster than consumption leading to land abandonment elsewhere), of neglect of land that was worse to work or access with heavy machinery, and of rapid and reckless urbanisation and industrialisation on the other. Both studied border regions differed from the interior particularly in a much faster transition of grasslands (especially meadows) to forests – or of arable land to grasslands and grasslands to forests. The decline of the share of grasslands was 7 pp in subset West Germany and 9 pp in subset

Austria, but only 2 pp in the interior; the differences in the growing shares of forest being similar. Another significant difference was a slower growth of built up areas in the regions adjoining Austria and especially West Germany when compared to the interior.

The second time period (1990–2000) represented a comeback for the market economy. In land use, the only significant trend was a loss of arable land to the detriment of grasslands, especially meadows. This change, inverse to the previous period to some extent, was a consequence of a sharp decline of agricultural production, and also of new state policies focused on grassing over of arable land, especially in less favourable and other environmentally sensitive areas. Not surprisingly, this change was much faster in both border regions. As can be seen from the figures, only about 1.5% of the total area of interior was over grassed during this period; but it was 4% of the area of the subset West Germany, and 3% in the case of subset Austria.

Therefore, these two border regions, when compared to the interior, are typical of a lower anthropogenic pressure on landscape, and of a steeper decline of this pressure – afforestation of grasslands during socialism, and grassing over of arable land later. Land use in subset West Germany is relatively homogenous, whereas the subset Austria consists of an intensively used eastern part and hilly western part, exposed to strong extensification.

All these differences can be attributed to several factors – both local ("proximate") and national ("underlying") (Mather, 2002). The most important local factors are probably worse natural conditions, subsequent lower levels of socio-economic activity/attractiveness, and expulsion of Czech Germans after 1945, followed by a non-perfect re-settlement (especially in highlands and mountains). These factors combined with growing intensity of agriculture and (sub)urbanisation in better conditions, and with underlying forces like increasing interconnectivity and regional specialisation within the whole system. All these forces resulted into marginalisation, land abandonment and extensification in the studied border regions – trends common to all less favoured areas in Europe (e.g. Kabrda, 2008 for Czechia; Sporrong, Ekstam, & Samuelsson, 1996 for Sweden; Krausmann et al., 2003 for Austria or Petek, 2002 for Slovenia).

The existence of the Iron Curtain as a barrier had definitely strengthened these trends in the socialist period – especially in its direct vicinity. But its influence should not be overestimated. The same changes, although weaker, would have taken place even if it had not been established, as can be seen by similar conditions along the borders with former East Germany, Poland and Slovakia (Bičík & Kabrda 2008).

6.4 Land-Cover Changes Along the Iron Curtain Using CORINE Data (1990–2000)

In the following, we will concentrate on land-cover changes along the former Iron Curtain between the years 1990 and 2000. As was described in previous text, land-cover information from the CORINE database was analysed in a belt 30 km wide,

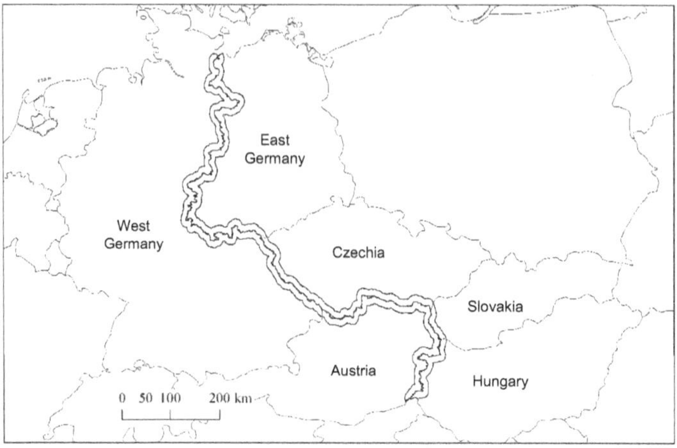

Fig. 6.7 Schematic map showing studied belt on both sides of the Iron Curtain (Source: own calculations, Najman, 2008)

stretching along the Iron Curtain (15 km on each side) from the Baltic Sea to former Yugoslavia (Fig. 6.7). Partial results are depicted in the tables (Table 6.2, 6.3 and 6.4); see Najman (2008) for more details. Several conclusions may be drawn from these data.

Table 6.2 (index of change) shows that land-cover changes in the studied period were much higher to the east from the Iron Curtain. The most intensive changes occurred on the Czech side of borders with former West Germany and Austria.

A mixture of political and economic reasons probably caused this difference. This territory could return to normal economic use once the restrictions connected with the Iron Curtain were abolished in 1990. Then, market forces began to influence the use of land, causing the landscape to return to a "normal" or "natural" course

Table 6.2 Index of change (IC, %, 1990–2000) in the belt along the Iron Curtain in different countries[3]

	Index of change (%)	
Section of the Iron Curtain	East	West
West Germany–East Germany	3.2	1.8
West Germany–Czechia	12.3	2.2
Austria–Czechia	9.3	0.2
Austria–Slovakia	7.4	0.2
Austria–Hungary	6.6	0.6
All observed territories	6.3	1.3

[a]Explanation: EAST denotes a belt 15 km wide to the east from the Iron Curtain; WEST denotes a belt 15 km wide to the west from it. See text for more explanation.

Table 6.3 Shares (%) of individual land-cover categories on both sides of the Iron Curtain in 2000, and changes in comparison with 1990[4]

	Share of category [%] in 2000			Decrease/increase (1990–2000)		
Category	Total	East	West	Total	East	West
Arable land	34.2	34.4	33.9	−4.6	−8.0	−1.0
Permanent cultures	9.9	5.8	13.9	−0.6	−1.4	−0.2
Grasslands	11.7	14.2	9.2	8.7	13.6	2.1
Forest land	36.3	36.9	35.7	1.2	2.7	0.2
Built up areas	4.5	4.6	4.5	5.2	7.1	3.5
Water areas	1.9	2.4	1.4	5.4	7.4	2.2
Other areas	1.5	1.7	1.4	−1.8	−6.6	4.6

Table 6.4 Land-cover changes (1990–2000) between individual categories on both sides of the Iron Curtain[5]

Category		Share (%) of the change of a sum of all changes		
1990	2000	Total	East	West
Arable land	Perm. cultures	3.4	2.4	7.8
Arable land	Grasslands	42.5	48.2	14.4
Arable land	Forest land	1.0	1.0	1.0
Arable land	Built up areas	4.6	3.5	10.0
Arable land	Other areas	1.1	0.4	4.4
Perm. cultures	Arable land	2.1	1.7	4.1
Perm. cultures	Grasslands	3.4	2.7	6.6
Perm. cultures	Built up areas	0.6	0.4	1.8
Grasslands	Arable land	6.0	5.7	7.3
Grasslands	Perm. cultures	1.4	0.8	4.1
Grasslands	Forest land	18.7	20.5	10.3
Forest land	Grasslands	10.4	8.0	22.4
Other areas	Water areas	1.6	1.9	0.2
All other changes		3.2	2.8	5.6
Total		100.0	100.0	100.0

[a]Explanation: each number denotes a share (%) of the given land-cover change of the sum of all changes; see text for more explanations.

of development. The eastern side of the belt along the Iron Curtain can be viewed as "frozen" or "conserved" between the years 1948 and 1990, trapped in the steady decay of socialism; consequently seeking to "catch up" with its western neighbour after 1990. In other words, the same development that took decades in the West must have been made in only a few years in the East.

Among minor, partial factors, may belong: (1) creation of new buildings and check points, especially on the eastern side of the former Iron Curtain; (2) reclassification of dead forests as grasslands in National park Šumava in 2000;

(3) creation of Dyje (Thaya) National park on the Czech–Austrian border; (4) development of new roads and highways in Slovakia near its capital of Bratislava, located directly on the observed belt; (5) the same applies to a new dam and regulations for the Danube river.

If we consider the nature of these changes (Table 6.3), the most important processes were the loss of arable land to the detriment of grasslands, forests and built up areas – changes more or less common for most European landscapes (see above). Significantly, the trends (direction of changes) were the same on both sides of the former Iron Curtain, but stronger to the east. This supports our previous statements – the East is "catching-up" with the West after a sudden change of political and institutional regimes (Jeleček, 2002).

Table 6.4, a summary of the transition matrix, provides information on real land-cover changes in pixels between the years 1990 and 2000. Differences can be identified between the eastern and western part of the studied territory. Only two processes were of real importance to the east of the Iron Curtain – grassing-over of arable land (almost 50% of all changes) and afforestation of grasslands (over 20% of changes). These trends, signs of extensification of land use resulting from renewed functioning of market forces and decline of agriculture after 1990, comply with the findings of the previous text. On the contrary, more types (even contradictory) of land-cover transitions occurred to the west of the Iron Curtain – the development was greater and smoother there. Besides the trends of extensification (grassing-over of arable land, afforestation of grasslands), also the processes of intensification (development on arable land, transformation of forests to grasslands) were important there.

We can summarize that all figures presented here document deeper land-cover changes on the eastern side of the former Iron Curtain. Furthermore, the same types of changes with the same intensity on both sides of the borders did not take place. In the eastern part of the studied territory, there were more common processes leading to extensification, connected with a loss of support for agricultural production. On the western side, a more regular distribution of changes among all possible types can be seen.

6.5 Conclusions

We have to stress that the two methods used in this article are not fully comparable. The first is based on cadastral statistics, describes land use, and has a certain degree of delay and inaccuracy in comparison with reality (especially because it is based on what the land owners/users report to Cadastral Offices). The second method, depicting land cover, and based on an interpretation of remote sensing data in the form of the CORINE database, has a relatively coarse resolution (minimal mapping unit) of 25 ha (5 ha for land-cover changes). Different methods and data sources can then lead to slightly different results.

For instance, cadastral data for the Czech borderland in the year 2000 display a lower share of permanent grasslands and a higher share of arable land when compared to the data from CORINE (see Najman, 2008 for details). Actually, a portion of the plots registered as "arable land" in the cadastre was covered with grasslands or in fallow in reality. The amount of unused arable land was rising until 2004 (when the Czech Republic joined the EU), finally reaching approximately 7–8% of the whole area of arable land in 2003 (about 300.000 ha – Zelená zpráva, 2003). Similarly, the CORINE database recognizes some tracts of dead forests in the Šumava mountains as grasslands, because of the high amount of green shrubs and grass below the dead trees.

However, using both methods led to similar conclusions within the territory and time period (Czech borderland in the years 1990 and 2000), implying a general ability of both datasets to capture the most significant trends occurring in the landscape.

Our research proved that regions along the former Iron Curtain, when compared to the interior, are typical of a lower anthropogenic pressure on landscape (less arable land and built up areas, more grasslands and forested areas), and of a steeper decline of this pressure (afforestation on grasslands, grassing over on arable land). Border regions act as "hot-spots" of land-use changes. As a result of general modernisation of "socio-economic metabolism", (Krausmann et al., 2003; Fischer-Kowalski & Haberl, 2007) border regions are being strongly extensified, taken out of traditional agricultural use, and transformed to satisfy other needs of modern society (nature and water protection, recreation, tourism). Thus, productive function, necessary in every locality in the era of closed local material and energetic cycles of the pre-industrial economy, is being replaced by non-productive functions in the era of open national or even global cycles of the industrial and post-industrial economy.

All figures presented here document deeper land-cover changes on the eastern side of the former Iron Curtain. Furthermore, the same types of changes with the same intensity did not take place on both sides of the borders. In the eastern part of the studied territory, there were more common processes leading to extensification, connected with a loss of support for agricultural production. On the western side, a more even distribution of changes among all possible types can be seen.

But the political border in the form of the former Iron Curtain was only one factor influencing differentiation of land use in the borderland. Other important factors were natural condition (soil productivity, altitude, slope, etc.) and also socio-economic characteristics (e.g., density of population, spatial exposedness) – in other words, functioning of the so-called "differential rent I" (Jeleček, 2002). Moreover, on the Czech side of the Iron Curtain, the expulsion of Czech Germans after WWII (ca. 1945–1947) and consequent non-perfect repopulation of these regions in the period of the totalitarian regime (1948–1989) had a large influence on the landscape in those regions. And this huge transfer of population and massive social and economic change was an important driving force of land-use changes even more than 40 years later – in the period after 1990. The behaviour of the new (relatively sparse) population, its problematic relationship to agriculture, farming, land and landscape were important factors leading to a large-scale land abandonment, grassing-over and afforestation. Furthermore, large parts of these relatively "empty" regions were designated as National Parks (Šumava and Podyjí/Thaya) by the Czech government.

Acknowledgements This work was supported by the Grant Agency of the Czech Republic, project no. GAČR 205/09/0995: "Regional differentiation and possible risks of land use as a reflection of functional changes of landscape in Czechia 1990–2010" (project leader: I. Bičík) and by the Ministry of Education, Youth and Sports, project no. MSM0021620831 "Geographical systems and risk processes in context of global changes and European integration" (project leader: L. Sýkora).

Notes

1. White colour marks the subset Interior; see text for more explanation.
2. "Sudetenland" consists of BTUs with more than 50% of population of German nationality according to the 1930 census; "peripheral spatial position" adopted from Hampl et al. (1987); see text for more explanation.
3. Index of change (IC, e.g. Bičík, 1995 or Jeleček, 2002) describes by one number the overall intensity of land-use change. The number, ranging from 0 to 100, shows the percentage of the whole area on which any land-use change occurred between the two time horizons. (Source: CORINE Land-cover database, own calculations, Najman, 2008)
4. Source: CORINE Land-cover database, own calculations, Najman (2008).
5. Source: CORINE Land-cover database, own calculations, Najman (2008).

References

Bičík, I. (1995). Possibilities of long term human-nature interaction analyses: The case of land-use changes in the Czech Republic. In I. G. Simmons & A. M. Manion (Eds.), *The changing nature of the people-environment relationship: Evidence from a variety of archives* (pp. 79–91). In Proceedings of the IGU Commission on historical monitoring of environmental changes meeting. Prague: Charles University.

Bičík, I., & Jeleček, L. (2003). Long term research of LUCC in Czechia 1845–2000. In L. Jeleček, et al. (Eds.), *Dealing with diversity* (pp. 224–231). Prague: Charles University.

Bičík, I., Jeleček, L., & Štěpánek, V. (2001). Land use changes and their societal driving forces in Czechia in 19th and 20th Centuries. *Land Use Policy*, *18*(1), 65–73.

Bičík, I., & Kabrda, J. (2008). Changing land use structure and its driving forces in border regions of Czechia. In J. Kabrda & I. Bičík (Eds.), *Man in the landscape across frontiers: Landscape and land use change in central European border regions* (pp. 33–47). CD-ROM Conference Proceedings of the IGU/LUCC Central Europe Conference 2007. In. Prague: Charles University.

Bičík, I., & Kupková, L. (2002). Long-term changes in land use in Czechia based on the quality of agricultural land. In I. Bičík, P. Chromý, V. Jančák, & H. Janů (Eds.), *Land use/land cover changes in the period of globalization* (pp. 31–43). In. Proceedings of the IGU-LUCC international conference Prague 2001. Prague: Charles University.

Bičík, I., & Štěpánek, V. (1994). Long-term and current tendencies in land-use: Case study of the Prague's environs and the Czech Sudetenland. *AUC – Geographica*, *29*(1), 47–66.

Chromý, P., & Rašín, R. (2009). Land use and land cover changes in Czech-Austrian borderland. *AUC – Geographica*, *44*(1), in print (19 pages).

EEA (European Environmental Agency). *CORINE Land Cover (CLC)*. Copenhagen.

Fischer-Kowalski, M., & Haberl, H. (Eds.). (2007). *Socioecological transition and global change: Trajectories of social metabolism and land use*. Cheltenham/Northampton: Institute of Social Ecology, Klagenfurt University, Vienna and Edward Elgar Publishing.

Hampl, M., Gardavský, V., & Kühnl, K. (1987). *Regional structure and development of the settlement system of the ČSR*. Prague: Charles University (in Czech).

Jeleček, L. (2002). Historical development of society and LUCC in Czechia 1800–2000: Major societal driving forces of land use changes. In I. Bičík, P. Chromý, V. Jančák, & H. Janů (Eds.), *Land use/land cover changes in the period of globalization* (pp. 44–57). In. Proceedings of the IGU-LUCC international conference Prague 2001. Prague: Charles University.

Kabrda, J. (2004). Influence of natural conditions on land use in the Vysočina region and its changes since the mid-19th century. *AUC – Geographica*, *39*(1), 15–38.

Kabrda, J. (2008). The changing spatial structure of agricultural land use in Czechia since the mid-19th century. *Geografický časopis*, *60*(3), 255–278.

Krausmann, F., Haberl, H., Schulz, N.B., Erb, K.-H., Darge, E., & Gaube, V. (2003). Land-use change and socio-economic metabolism in Austria – Part I: Driving forces of land-use change 1950–1995. *Land Use Policy*, *20*(1), 1–20.

Lambin, E., & Geist, H. (2007). Causes of land-use and land-cover change. In C. J. Cleveland (Ed.), *Encyclopedia of Earth*. Washington, DC: Environmental Information Coalition, National Council for Science and the Environment.

LUCC UK Database. – database of the project of the Grant Agency of the Czech Republic GAČR 205/09/0995: *"Regional differentiation and possible risks of land use as a reflection of functional changes of landscape in Czechia 1990–2010"*, Charles University, Prague, http://lucc.ic.cz

Mather, A. S. (2002). The reversal of land-use trends: The beginning of the reforestation of Europe. In I. Bičík, P. Chromý, V. Jančák, & H. Janů (Eds.), *Land use/land cover changes in the period of globalization* (pp. 23–30). In Proceedings of the IGU-LUCC international conference Prague 2001. Prague: Charles University.

Najman, J. (2008). *Evaluating influence of the Iron Curtain on landscape changes using the CORINE land cover data*. Master thesis, Charles University, Prague (in Czech).

Petek, F. (2002). Methodology of evaluation of changes in land use in Slovenia between 1896 and 1999. *Geografski sbornik – Acta Geographica*, *42*(1), 61–97.

Sporrong, U., Ekstam, U., & Samuelsson, K. (1996). *Swedish landscapes*. Stockholm: Swedish Environmental Protection Agency.

Štěpánek, V. (2002). Czech frontier in the 20th century: Major political shifts reflected in changing land use structure. In I. Bičík, P. Chromý, V. Jančák, & H. Janů (Eds.), *Land use/land cover changes in the period of globalization* (pp. 110–115). In Proceedings of the IGU-LUCC international conference Prague 2001. Prague: Charles University.

Zelená zpráva. (2003). *(Green report 2003). The report on the state of Czech agriculture in 2003*, Ministry of agriculture of the Czech Republic, Prague (in Czech).

Chapter 7
Landscape Function Transformations with Relation to Land-Use Changes

Ivan Bičík, Jiří Anděl, and Martin Balej

7.1 Land-Use Analyses and Landscape Assessment

Land-use changes at different scales (from global to local) are a key topic in a number of scientific branches, such as geography, sociology, economy, landscape ecology, and others (Himiyama, 2002; Leppers, 2002; Naveh, 1991; Worster, 1985; Žigrai, 1996). This theme has much to do with the basic challenges brought by the dynamic changes of modern society. Such challenges include nature–society interactions, the search for natural resources that would secure adequate nutrition, population rise, connections between land use and climatic changes and many other pressing problems (Himiyama, Mather, Bičík, & Milanova, 2005; Turner II et al. 1990).

The study of land-use development in localities and regions gives us the possibility to understand the most recent developments in interaction between Nature and Society and to research also the main driving factors influencing directions and intensity of changes in the landscape. Development of the interrelation between Nature and Society started in the pre-industrial period. Hampl (1998) described this phase as a period of determination (distinctive dependence of society on natural conditions), distinguished by a dominance of residential and productive landscape functions. The characteristics of the landscape determined its function up to a certain point. The industrial period followed. Society determines the function of the landscape and at the same time becomes an important, even fundamental, factor that competes with natural conditions (Hampl's phase of competition). Society also puts pressure on a different land-use structure. In the post-industrial period, the interrelation between Nature and Society achieves a cooperative tendency (Haines-Young & Potschin, 2003, according to Hampl 1998, the phase of cooperation), so far applied only in some of the most advanced societies. Multi-functional land use and sustainable trajectory of landscape development have been sought (Naveh, 2007). Only on

---

I. Bičík (✉)
Department of Social Geography and Regional Development, Charles University in Prague, Albertov 6, 128 43 Praha 2, Czech Republic
e-mail: bicik@natur.cuni.cz

J. Anděl et al. (eds.), *Landscape Modelling*, Urban and Landscape Perspectives 8,
DOI 10.1007/978-90-481-3052-8_7,

this base can we predict future possible forces which may influence development of land use in landscape of specific functions. There are also new influences which play a role. Globalisation and its impact on effectiveness of farming in different conditions, brings changes in intensity of farming and in land-use structure (Douglas, Huggett, & Robinson, 1999; Haberl, Batterbury, Moran, & 2001; Haberl et al., 2002a, 2002b).

Current remote-sensing techniques allow global-scale monitoring of land use and land cover, as well as computer-assisted creation of land-use and land-cover maps (Lambin, 2002; Lillesand & Kiefer, 2002). In order to understand the current state of land use, and to investigate future trends, it is especially important to have a sound knowledge of past trends (Antrop, 2003, 2005). Historical land-use studies, however, often lack reliable information sources. Therefore, any data containing historical land-use information are very valuable, regardless of the scale (Meyer & Turner, 1996). If historical land-use data are relatively complete and allow time analyses, it is possible to search for "driving forces" that crucially influenced past land-use changes and the interconnections between land use and socio-economic conditions.

Regional land-use studies make it possible:

- To compare the importance and structure of the "driving forces" at different scales – global, continental, national, regional and local. Global-scale land-use analyses have special importance; moreover, such research also brings theoretical contributions as geography as a science is supposed to examine how similar concepts may show different results on different scales and under different conditions.
- To analyse the extent of different land-use types and its structure in different regions in the light of uneven regional conditions. As basic social and economic conditions in regions/states differ from each other, historical land-use analyses (provided they reveal the importance of "driving forces") allow us to predict future land-use trends.
- To verify theoretical results of land-use analyses and their regional variations in physical and regional planning. Thus, historical land use forms a practical base that helps to determine the desirable functions of different planning regions and also contributes to the policy-making process.

Special orientation in our topic shows changes of functions in different regions and influences on changes in their land-use structure. Post-communist countries present a special opportunity to study the influence of a political and economic system on land use in different regions (Palang et al., 2006; Skowronek, Krukowska, Swieca, & Tucki, 2005; Bičík & Jančák, 2002). Loss of a centrally planned economy in these countries brought special factors influencing land-use changes: the loss of a closed economy and impact of global prices especially in agricultural products; restitution of property; reorganisation of socialist cooperatives in agriculture into cooperatives of owners; privatisation of state farms; significant strong process of suburbanisation; new impulse for regional and supra-regional connections –

highways, railway main lines, higher care for the environment, etc. These changes in agricultural and rural space were profound. The number of agricultural land users during the totalitarian period was some hundreds of socialist farms. Land was owned by the state and was given to cooperatives and state farms for their collective use. Czechia has now, after an almost-finished process of restitution, some 3.5 million land owners!

7.2 Case Study – Northwestern Part of Bohemia

We document typical changes in land use of different types of landscape in the Czech–German border area (northwestern part of the *Sudeten* lands) from the mid-Nineteenth century on three sample areas (Fig. 7.1). They are distinguished by dissimilar natural conditions, economical-geographical location, and socio-economic structure. This region of Czechia stretches along the border with Saxony, and was populated mainly by Germans (over 90% of the total population) until the middle of the Twentieth century. The state border was stable, but the ethnic border between the Czech and German population had been changing frequently. Since the Thirteenth century, it had been changing to the disadvantage of Czechs, especially after the 30-years' war, when the loss in Czech population was compensated for by a German-speaking population. Since then, the orientation of economic, cultural and political ties has also been changing. The ties were much stronger with cities in Saxony than with cities in the Czech inland (even after the emergence of Czechoslovakia in the year 1918). Commonly, borderline cities with their mainly Czech-German population were much richer than cities in the Czech inland with their mainly Czech population. The distinctiveness of the *Sudeten* region is reflected, among other things, in a different character of landscape, distinctive folk art and architecture (cf. Raška, 2006; Oršulák, Raška, & Suchevič, 2007). For example,

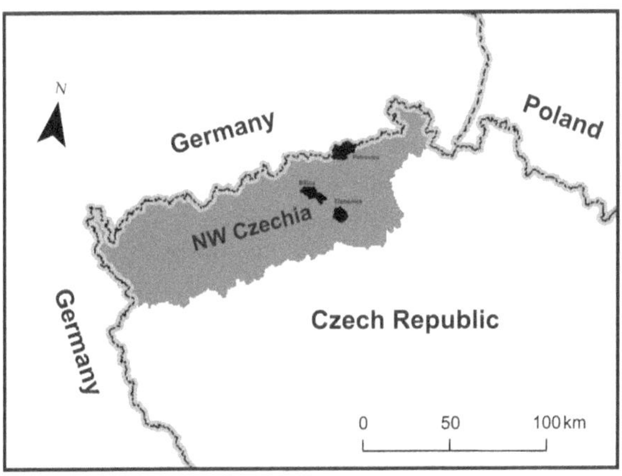

Fig. 7.1 Study areas Bílina, Petrovice and Třebenice within the Northwestern part of Bohemia (Czech Republic)

even the breed of cattle that they raised was different. That's why in connection with historical development and characteristics of settlement, the term "*Sudeten* landscape sample" is occasionally used.

Tens or hundreds of small workshops and factories which carried on the old tradition of home production (canvas, glass, woodwork) were strongly scattered regionally, and created a distinguishing characteristic of the landscape. In exceptional conditions, some areas with small metalworks and manufactories developed into strong industrial centres (for instance, Děčín or Chomutov). The landscape was interwoven with a dense network of trails, which usually reached dominants located on heights (towers, lookouts or guesthouses with restaurants). Tourists took advantage of spatially dispersed accommodation in summer homes or family pensions, so there was no concentration in recreational resorts as there is now. The landscape was burdened pretty evenly, and a dense railroad and road network enabled people to commute to work in factories. Communication pathways were distinguished by many small architectonic elements (overpasses, footbridges), which were delicately placed into the landscape. Waterways were often edged with walkways. Use of energy from waterways was also unique (small water power stations).

Gradual concentration of population and economic activities, together with landscape use *restructuralisation* (induced by industrialisation and urbanisation and emphasised in the totalitarian period 1948–1989) led to formation of new requirements on the landscape. The landscape changed from having a predominantly productive (or residential-productive) function to a multi-functional one. Larger regional units were gradually created, with different dominating functions: core regions (with industrial, mining, residential service industry); and transitional and peripheral regions with functions that are less disruptive for the environment (extensive agriculture, mainly concentrated on pastoral cattle husbandry, ecological and water management functions, etc.).

Type A: "Coal basin" Bílina area (total size 4.600 ha, average altitude 200 m above sea level), is placed in a favourable and exposed location in Most basin on the regional development axis (Cheb – Karlovy Vary – Chomutov – Ústí nad Labem). This rolling, and now practically deforested landscape lies at about 200 m above sea level. Relief in the northern and eastern part of this area is significantly anthropogenically affected because of extensive open-pit mining activities. A historical trading road stretched along the Bílina River catchment, which flows through the central Bílina area. Bílina was inhabited in primeval times, as archaeological remains in today's Bílina town document. There was a Slavonic fortified settlement here in the Tenth century. This previously very fertile and intensively used agricultural region developed into a region with quite new functions, because of the development of large-scale open-pit brown coal mining. These functions are "dictated" not by local reasons but by whole-society requirements. Many small settlements have perished. Population differentiation is distinguished by extreme polarisation and concentration of population in one location. Currently there is very high employment in industry (43%), and conversely very low employment in agriculture. Of 15.900 inhabitants, only 51% were born here (so they are "natives").

Type B: "mountain periphery", Petrovice area (total size 6.400 ha, average altitude 650 m above sea level), was in a very peripheral location at the eastern part of Krušné hory (mountains) after the Second World War. Its relative geographical position significantly improved after 1989, when highway D8 was opened on the international transport axis Prague–Berlin. The tableland is mostly covered with forest and cultivated meadows. From the east, the sandstone outcrops of Tiské stěny (a protected-landscape area) extend into this tableland, and are a tourist attraction. The dominant of the northwestern part of this sample area is a hill, Špičák (723 m above sea level). There are abundant sources of underground water. Water is channelled into a water pipe system which supplies conurbations in Krušné hory (mountain) foothills. The low density of residences corresponds to an extremely low density of population (only 20 inhabitants per square km). The Petrovice area is distinguished by a lower number of "natives" (36% only), extremely low proportion of industrial employees (20%), and a high proportion of houses used primarily for recreational activities (28%). This area represents analogous regions in Krušné hory (mountains), which show dynamic development already in the first stages of industrialisation (in direct connection with the development of Saxon cities). Stagnation of economic and demographic development follows. After the Second World War, this area also shows significantly regressive trends and total change of landscape function and character.

Type C: "Intensive agricultural area" – Třebenice area (total size 3.600 ha) with predominantly flat lowland and sparsely wooded relief, represents landscape intensively used for agriculture, with alternating plots of fields, orchards and tiny stands of trees. From the north, the Třebenice area is penetrated with steep slopes of the volcanic-sedimentary complex of *České středohoří* (Czech Middle Mts.), a landscape-protected area. The warm and dry climate at an altitude about 250 m above sea level, together with the potential of fertile black soil, creates the preconditions for intensive agricultural use. The road from Louny to Lovosice is the axis of the Třebenice area, and connects it with highway D8. The little town of Třebenice, which is located on an old trade road, became a centre of gravity. The Třebenice area is distinguished by a higher population of older inhabitants (19.2% of 3.800 inhabitants are older than 60 years); an above-average proportion, within northwestern Bohemia, of "natives" (49%), a higher proportion of workers in agriculture (9%) and a relatively low proportion of employees in industry (25%). Lack of work opportunities in the region necessitates a high level of commuting to work places (65% of economically active workers).[1]

7.3 Assessing the Land-Use Changes

We assess land-use changes in four time horizons. They have specific characteristics in the Sudeten region, and correspond to phases of development of society (Hampl 1998) which are as follows.

The *pre-industrial* phase, which is characterised by dominant employment in the primary sector – agriculture (with ore extraction, forestry in mountainous areas), and low developmental dynamics. Natural determination with a low spatial mobility plays an important role for distribution of inhabitants. Population density is relatively even, and it ranges in sample areas in the year 1850 from 75 inhabitants (Chudenice by Bílina) to 250 per square km (Bílina). A limited role of cities in the settlement system leads to a low degree of hierarchic organisation. In comparison to the following phases, population is distributed relatively evenly, and individual settlements have low size range (100 inhabitants in Chudenice by Bílina and 3,700 in Bílina in the year 1850).

The *industrial* society is characterised by development of the secondary sector and strong dynamics. Natural determination is being overcome, and the role of social geographical factors keeps increasing. The process of urbanisation is pressed for. This process is connected to high spatial mobility. This form of urbanisation is described as extensive, and develops first in northern Czechia. There is a strong bond to Saxon cities there. The process of industrialisation also starts in the Czecho-Saxon borderlands. It spreads from there to other parts of northwestern Sudeten, meaning from east to west, and from there, further into the Czech inland. The range in population density is increased dynamically, from 60 to 750 inhabitants per square km in the year 1921 (in Krásný Les by Petrovice and Bílina, respectively).

The *totalitarian* period (German occupation and communist era) represents the final stage of industrial society and diversion from the natural trajectory of advanced Europe (which was already showing certain features of a post-industrial society). For northwest Czechia this era means a period of reversal in development and interruption of existing developmental continuity. It is foreshadowed by resettlement of residents of German nationality. Vast disruption of settlement structure followed, liquidation of housing stock and destruction of many historical and artistic monuments. At the same time, there was enfeeblement of identification of "newcomers" with the landscape. This was reflected in perishing of many local customs and traditions. Newcomers were not "at home" here. This situation has been changing only with the second and mainly the third generation. The ties with cities in Saxony were significantly subdued. Large capacity, open-pit brown coal mining is developing in Krušné hory (mountain) foothills and heavy industry (mainly energy and chemical) is connected to mining. Large-scale agricultural production pushes through in fertile areas. Strong polarising trends culminate in organisation of settlement structure. In spite of proclamation of tendency of equalisation, the asymmetry, core x periphery (hinterland) is the strongest. Differentiation in population density is deepening, from 5 to 1,000 inhabitants per square km in the year 1980 (in Krásný Les and Bílina, respectively).

The *post-industrial* society, with its developing tertiary sector, is characterised by intensive development of communication (pressure on the transport system) and informational contacts (Internet, digital phones). In settlement structure, it is coming to an integration of system. Depopulation developmental trends are pushing through at big cores on the account of space in their hinterland. Changes of geopositional conditions lead to pushing through of former peripheral areas along the border with Germany. Tourism (*recreational industry*) represents for many of them the main carrier of economic and social rehabilitation of an area (Petrovice area).

7.3.1 Data Sources and Their Origins in Czechia

Each of the monitored periods ends with a time horizon. For this time horizon, there are at disposal land-use data and census data of population and houses of respective sample areas – 1845, 1948, 1990 and 2000. Evaluation of functional changes of the landscape as a key factor for following changes of land use requires adequate data and an adequate methodological basis. There is not too much data about long-term changes of land use, especially for a period longer than a century. GIS subsequently offers very strong support for data processing and further modification. The ability to synthesize information about spatial phenomena with the help of integration of georeferential data enables researchers to generate quite new information (Feranec & Oťahel, 1992, 2000). Creating a geographical database, as described by Jensen (2005), today constitutes one of the main research phases of different scientific branches (not only of geoscience).

Detailed land-use data were first collected on the Czech territory more than 180 years ago as part of cadastral records (originally called *stabilní katastr – stable cadaster*). In the early Nineteenth century, it was necessary to gather data that would serve as a base for land tax calculation (which was at that time the main source of the state budget). As a side effect, a precise triangulation network came into existence and cadastral maps (scale 1:2.800) were created. These maps were later transformed to the scale 1:2.000. The actual plot sizes were calculated from these cadastral maps and all maps of smaller scales were derived from them (Mašek, 1948).

Preparatory work started in 1816. Taxation was based on plots and net income from each plot has been calculated. One map (composed from a number of sheets) with precise plot boundaries was drawn for each cadastral unit. Detailed mapping was carried out between 1826 and 1843; it included almost 13.000 cadastral units with more than 15 million plots in Bohemia, Moravia, and the Czech part of Silesia. The images covered almost 50.000 map sheets. Finally, plot areas were calculated and all plots were evaluated on the grounds of land use and land cover. The structure of land by cadastral units was also recorded; altogether 52 land-use and land-cover types were recognised. These records date back to 1845 and have survived in archives.

There have been many changes in the cadastral records over the decades. Later revisions brought new cadastral units and new land summaries into existence. Simplified land-use data for more than 13.000 cadastral units were summarised in 1948 by the Central Survey and Cadastral Archive in Prague. Data for 1990 and 2000 originate from the computer database of the Czech Land Survey Office (Cadastral bureau).

7.3.2 Maps

Utilisation of maps from cadastral mapping for evaluation of long-term landscape changes from 1826 to 1843, runs into the fact that mapping was done at a detailed scale of 1:2.880 for 13.000 cadastral units. The total number of map sheets reaches about 35.000, and it is obvious that their use in research is therefore somewhat complicated in any effort to analyze a larger area. The set of map sheets has now been digitised, and represents an exceptional database for evaluating landscape changes. The basic procedure which was used in different sample areas is based on: (1) Comparison with maps of different time horizons in GIS. (2) Evaluation of range changes of individual categories and their layout in a cadastre unit. (3) Evaluation of stable and instable areas, and finally (4) on total intensity of changes (index of change, coefficient of ecological stability, coefficient of anthropogenic effect and the like (for example Bičík, 1997; Bičík et al., 1996 and others). The procedure is quite demanding. It requires content and statistical unification of database information, and obviously the generalisation of outputs is rather difficult. Without any doubt, this procedure can serve as an interesting base for putting into practise detailed evaluation of landscape changes for specific purposes (complex land reform, marking out local and regional territorial systems of ecological stability (*ÚSES*) and the like). The procedure is more for application than for research. Only the evaluation of several tens of sample areas, made of approximately five to ten cadastral units in different positional, natural, and functional conditions could be a path for scientific research. Creation of an adequate generalised map, for instance in scale 1:250.000 for entire Bohemia is methodically extremely difficult due to the necessary generalisation.

7.3.3 Remote Sensing Data as a Comparison Dataset

The importance of the remote sensing method has been confirmed by its own rapid development (Lillesand & Kiefer, 2002; Walsh, Evans, & Turner, 2004). The remote sensing data is, however, limited to the period from the middle of the Twentieth century until today; the mass research use of photography in the process of mapping the landscape did not develop until after the end of World War II. A very important source of information for observing the state of the landscape from 1945 is historical aerial survey photographs (Paine & Kiser, 2003). Three sets of aerial photographs of the study areas from 1948 have been obtained. Contemporary land use of the study areas is accurately documented by digital orthophotomaps that have their origin in the ORTO ČR project with the pixel size of 50 cm for the target scale of 1:5.000; these data date back to 2004. The bases for the colour orthophotomap were aerial survey photographs. They were digitised and provided in the uncompressed TIFF format; georeference is in TFW format in the S-JTSK coordinate system (Unified Cadastral Trigonometric Network).

The input data had to be adjusted to correspond to tasks, objectives and specific features of landscape ecological investigations. Aerial photographs were digitised using the photogrammetric scanner DSW200 Helava into the TIFF format with resolution of 1.500 dpi. The orthophotomap was created by scanning the aerial photographs with accuracy of 14 μm, which is approximately equal to 1.800 dpi (dots per inch) and corresponds to the pixel size of 50 cm. A georeferencing and orthorectification of the aerial photographs took place within the Erdas Imagine and ArcView 9.2 programs (for reading information in GIS).

7.4 Land-Use Changes and Transformations of Landscape Function

The Bílina area is an area where the function of the landscape completely changed during its development. It changed from originally purely agricultural (one of the most fertile regions in Bohemia), to industrial-agricultural, and eventually as far as urban and totally devastated landscape. Immediate negative anthropogenic impact affects the area northwest of Bílina city. The area had already been mined for coal since the Eighteenth century. Large capacity pit mining extended formerly local mining from the 1960s, uncovered the surface of the landscape, and created vast anthropogenic forms of topography (mining pits and tips). In the 1970s, three municipalities vanished. There was a total population of 4.000 people there before World War Two. A similar situation was present southeast of Bílina city, where Radovesice tip was created. Shortly after the Second World War, Radovesice had the character of a small town with more than 1.400 inhabitants. In the 1970s, earth was taken to the location of the old mines, and the municipality Radovesice was buried under a huge elevation. These changes caused a decrease in the proportion of agricultural land from 70% before the Second World War to the current 10%. Conversely, other areas (mainly mining sites) increased from 4 to 65% (Fig. 7.2). Modern construction and transformation of the landscape in the last few decades has completely changed the character of the landscape. Extensive industrial and large-scale, high-rise apartment-building construction, together with artificial forms of relief, overpower the predominantly original natural and urban environment. All these also cause a negative perception of the landscape, in spite of the developing restoration work on the Radovesická tip. Perhaps only the renovated historical centre of Bílina city, with a town square based on the medieval ground plan with a towering castle above, could be described as a positive cultural-historical "oasis" of the Bílina area. This green island is the site of Kyselka, a former spa (see typology, rate of concentration and development of population).

The densely forested mountain region of the Petrovice area was settled in the Twelfth century in connection with mineral extraction. The first settlers came from the Czech inland; later, immigrants from Germany dominated (mainly miners with their families). After the decline of mining, a mainly self-sufficient type of agriculture and logging started to push through. The largest landscape character change connects to events after the Second World War. After the post-war resettlement of the German population, the area remained permanently underpopulated. The population dropped to one quarter, and settlement structure significantly changed. Of ten original settlements, only three remained. During the totalitarian period, political power was also strongly pushing through. This meant waves of collectivisation and "new" organisation of the countryside. Former tiny fields and meadows were united under the terms of founding of Unified Agricultural Cooperatives (JZD) or used for large-scale cattle husbandry. Partial transition from vegetable production to animal farming, or deserting locations unsuitable for agricultural activities, is reflected in a change of land-use categories. There is a decrease of arable land (from 49 to 31%), in favour of forest (from 28 to 35%), meadows and pastures (from 18 to 24%, Fig. 7.2). The 1990s meant the disappearance of employment possibilities in collective farms or in small industrial companies for the Petrovice area. Nevertheless, this period provided an impulse for development of entrepreneurship, especially in tourist industry services. (That's why arable land decreases from 30% in 1990 to the current 13%; Fig. 7.2). This phenomenon shows itself particularly in the eastern part of the region.

The Třebenice area is intensively used for agriculture, and through its development shows stable structure of land-use categories (Fig. 7.2). The landscape character of the Třebenice area and its function were affected the most by a wave of collectivisation in the second half of the Twentieth century, in the totalitarian period. During this epoch, tiny plots of arable land, cultivated until then by private farmers, were united into vast fields. This collectivisation was ordered by state organs. So the face of the landscape significantly changed. Many ridges, stands of trees, woody accompaniments of streams and paths, "solitary" and specific alluvial plain communities almost disappeared from the landscape. Typical agricultural vernacular architecture, church buildings and minor composing landscape elements were in most cases destroyed. Most buildings in the Třebenice area are newly renovated for

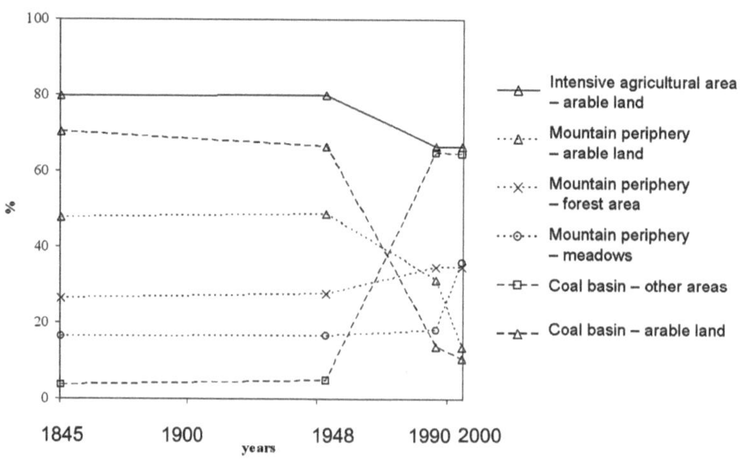

Fig. 7.2 Land-use changes in the study areas; using selected land-use categories with the most dynamic changes

housing and recreation. And this, up to a certain point, causes the disappearance of an original, specifically agricultural character of most local settlements. Green vegetation has been planted recently in continuously built up areas of municipalities, and so today's village centres make a clean and pleasant impression.

7.5 Interpreting the Driving Forces of Changes

The most dynamic changes in northwestern Bohemia occurred after the Second World War. The majority of the landscape in this region lost its agricultural function. The reason for that was loss of landholders (especially family farms). There was a drastic experiment in the *Sudeten* area. Original inhabitants were replaced with newcomers who didn't have any knowledge of tradition and links, and had a completely different attitude to their new possession. The experiment confronts landscape and its memory with newcomers (Balej & Anděl, 2008). At the same time, inhabitants/landlords were replaced by inhabitants/consumers. As a result, the *Sudeten* region became a memento, which shows what could happen on a much larger scale if it comes to a significant disruption of natural development of landscape (Lipský, 1994). The *Sudeten* region was strongly disrupted. Firm points in the landscape, cart tracks, thousands of stone statues, wayside shrines, tiny chapels and memorable trees disappeared (Lipský, 1996). Collectivisation meant, as a result, the change of a charming landscape into an anonymous expanse of fields. How should the landscape in northwest Czechia develop and in which direction should it go? There are several possibilities: to use existing settlement structure intensively, which should be unambiguously defined to a clear, non-built up landscape. To emphasize natural dominants of settlements (suppress chaos), to restore spatial and meaningful hierarchy of settlements in the landscape. To restore a network of historical footways, and connect those to whatever valuables in the landscape are still left. Footways should serve for journeys with some destination. A responsible landlord, who will work on and with the land, not only exploit it, should be found for each landscape site.

On the basis of elementary developmental trends in combination with geopositional factors, very simplified typology can be done (Table 7.1): the 1st type, "Coal basin", lies in a very exposed location in the basin under Krušné Hory Mts. During the totalitarian period, a dynamic change of landscape is characteristic for this type, connected to liquidation of settlements, and concentration of population to core areas. Also, a completely dominant transformation of elementary landscape functions, gradually from an agricultural to industry-agricultural, and in the totalitarian era to an urban and mining one is apparent. Regionally exposed areas, where mining and urban functions of landscape predominate, are distinguished by dynamic land-use changes, extremely high concentration of population, and huge interference with the environment of inhabitants.

The 2nd type, "Mountain semiperiphery", was distinguished in the totalitarian era by strongly peripheral features. This type remained underpopulated after

Table 7.1 Basic characteristics of sample areas[2]

Indicator/Type	"Coal basin"	"Mountain periphery"	"Intensive agriculture"
Geoposition	Exposed high profile, prominent	Semiperiphery → semiperiphery → periphery → semiperiphery	Semiperiphery
Location (a.s.l., in m)	Basin (lowland) 195–215	Mountainous 450–750	Lowland 170–260
Year of maximum population	1930	1880	1910
Decrease of population 1930–1950	by 30%	by 72%	by 25%
Disappearance of settlements after the year 1945	Liquidation by coal mining (seven settlements)	Disappearance of six settlements as a result of undersettlement	0
Number of settlements[a]	8 →8→8→ 1→1	9→9→9→3→3	11→11→ 11→11→11
Average settlement size[a]	670→2,670→ 2,020→17,025→ 15,700	890→770→ 480→400→430	370→510→380 →340→350
Rate of concentration (% inhab. in centre)[a]	69→69→73→ 100→100	33→39→37→ 41→39	34→36→35→ 37→35
Land-use type	Dynamic changes	Gradual changes	Relatively stable
Period of essential changes	Totalitarian	Totalitarian and postindustrial	Totalitarian
Basic structural land-use changes	Dynamic decrease of arable land in favour of other areas	Decrease of arable land in favour of meadows and forest areas	Mild decrease of arable land
Basic function	Agricultural→agro-industrial→urban and mining	Forest and agricultural→ recreational	Agricultural
Example	Bilina area	Petrovice area	Třebenice area

[a]Describes five time horizons – 1850, 1930, 1950, 1990 and 2000; the year 1930 represents a population maximum for the majority of settlements.

resettlement of the German-speaking population, (decrease of population by 72%). This decrease showed itself negatively by disruption of the structure of settlements (elimination of 2/3 of settlements), and by dynamic decrease of arable land as a result of lack of manpower. These tendencies continue also in the post-industrial period. In this last period, the agricultural function is replaced by a recreational function.

The 3rd type, "Intensive agriculture", represents a stable type with stable settlement structure and relatively low decrease of population after the Second World War. This type is characterised by only mild decrease of arable land in the totalitarian era. An agricultural function persists during the entire monitored period, even if in the post-industrial phase other functions are also pushing through, especially recreational ones.

The 1st type represents a polycentric core in the Krušné hory mountain basin areas, where the energy industry and mining function are dominant. The 2nd type represents the tableland of the Krušné hory mountains, a mostly desolate, marginal area. After German-nationality population resettlement, it has been permanently undersettled by people with weakened historical ties to the landscape. The 3rd type is typical for an intensively used agricultural area with high soil quality, and suitable for crop production.

7.6 Placing Monitored Issues into Wider Spatial and Time Scales

Altogether, a Czech land-use scientist can make use of an extensive set of land-use data hardly to be found anywhere else in the world. It contains the size of land-use categories in all cadastral units. For the sake of historical comparisons, so-called Basic Territorial Units (BTUs) have been created; these contain data from 1845, 1945, 1990 and 2000. There are some 9.000 BTUs; their average size is ca. 8 km^2. Approximately 70% of them consist of one cadastral unit only; the rest are composed of two or more, so that BTUs would not differ in total area by more than 1% in all years observed.

The character of data allows us to analyse the current state, past development and future prospects of individual land-use categories, as well as the total land-use structure at local, regional and national levels (Table 7.2). For details about methodology, see Bičík (1997, and others).

Table 7.2 Typology of changes of the macrostructure in BTUs of Czechia (Source: LUCC UK Prague)

	1845–1948		1948–1990		1990–2000	
	% of BTUs,	% of area	% of BTUs,	% of area	% of BTUs,	% of area
− − −	0.10	0.24	0.00	0.00	0.45	0.45
− − +	10.66	13.21	8.46	9.56	13.65	17.73
− + −	14.34	13.14	0.53	0.42	9.45	11.49
− + +	45.57	46.47	90.79	89.95	31.78	32.39
+ − −	7.97	7.65	0.01	0.00	8.71	8.19
+ − +	17.96	16.29	0.12	0.05	5.90	5.52
+ + −	3.06	2.55	0.07	0.01	24.58	21.14
+ + +	0.34	0.45	0.02	0.00	5.48	3.09
	$N = 8.910$	Area = 78.868	$N = 8.910$	Area = 78.868	$N = 8.910$	Area = 78.868

Explanation: In column 1, there are three signs. First sign = size of agricultural land, second one = size of forest areas, third one = size of "other areas" (built up + water + remaining).[3]

Remarks:

- Every period is characterised by completely different land-use development.
- While in the first period, types where an increase of the arable land had occurred (most often in combination with an increase of permanent cultures and optionally meadows) asserted themselves in more than 50% of the territorial units, the types with an increase of the area of arable land in the second period occurred in less than 15% of the units.
- In the period 1948–1990, the types with a decrease of the arable land and with an increase of permanent crops and optionally of meadows, dominated.
- Latest development – the period 1990–2000 is characterised by types with an increase of meadows, permanent cultures and pastures and with a decrease of arable land. It is essential to emphasize that in comparison with the previous period, this time, where the types of development characterised by an increase of the area of arable land occur (more than 35% of the units and more than 30% of the area), the number of units significantly increased.

Before the data evaluation of a detailed land-cover analysis of sample areas, we have to point out several specifying facts. First of all, sample areas defined from RS (remote sensing) and according to BTU are slightly different, namely in five BTU. Their definition, based on RS is only part of the area. Secondly, there is a specific fact, coming from the method used, which evaluates landscape changes with the help of typology of areas' macrostructure. The fact is, we evaluate zero change in range AGL (agricultural land), forest or other areas the same as growth, which means sign +. In the monitored complex, this fact appears four times. Furthermore, there is an impact of zero presence of forest areas in these four BTUs.

7.7 Results of Evaluation of Sample Areas

The development of land-cover macrostructure was, in all three monitored regions, strongly differentiated. In the first and second period, type – + + characterised sample areas Bílina and Třebenice, which means *ZPF* decrease, and growth of forest and other areas. In sample area Petrovice, the development of macrostructure was different in the first and second period: in the first period it was type + + –, and in the second – – +. In the third transformative period, functions of these three sample areas changed, and the result is a different development of macrostructure of land use.

It is significant that the above-mentioned characteristic of development of land-cover macrostructure in sample areas as a unit, is a result of rather opposing trends from a BTU point of view, by which sample areas are formed. Only the smaller part of the BTU has an identical type of development of land-cover macrostructure in comparison with sample land cover as a unit.

Overall, the largest changes in land-cover structures (be they followed from the point of view of macrostructure – three categories, or complete structure with eight categories) were realised during the years 1948–1990. From a total of 21 BTU in three monitored sample areas, 13 of them reached changes in at least one category by more than 50% of its dimension in the previous time horizon. This occurred in all four BTUs in the Bílina area, in five BTUs in the Petrovice area and in four BTUs in the Třebenice area! In the oldest period of similar fundamental changes, such changes occurred only in four BTUs; then, in the period 1990–2000, only in two BTUs in the Bílina area. So we can conclude that the process of landscape changes in Czechia was by far the most intense during the communist regime (see for example Bičík & Jančák, 2002). It was also similar in all three monitored sample areas.

It is characteristic that there were fundamental changes of landscape functions in sample areas in individual BTUs several times. After all, only in four BTUs in all three monitored periods is there the same type of change of land-cover macrostructure. It is always type – + +, which is the most frequent trend in the entire BTU set in Czechia (portion in individual periods of this type in Czechia was 46.5, 90, 32.4%).

Should we attempt to summarise the evaluation of changes in land-cover macrostructure, it is necessary to emphasise, that the different development in the three sample areas from the point of view of land-cover macrostructures, has been successfully proven. Furthermore, this differentiated development was characterised also by a different intensity of this development. These changes are the result of economical-social development of the entire republic. The impact on sample areas is most of all different agricultural function in all three sample areas. Also, profound were the changes in coal-mining extent and industry development mainly in the Bílina area, and related population concentration in this region. Finally yet importantly, the trend of the last 18 years also became evident. After long discussions, the agrarian policy of Czechia was accepted. It supports agriculture and countryside (both EU and national programs) naturally in a multi-functional country landscape (for more, see Bičík, Chromý, Jančák, & Janů 2002). And it is also evident in researched sample areas.

Acknowledgements The research presented was supported by the project *Czech Borderland after Schengen: a Distinct, Oscillating and/or Transit Area?* (No. IAA311230901) supported by the Grant Agency of the Czech Republic.

Notes

1. For comparison, the ratio of agricultural workers (2001) was, in Ústí nad Labem district 3.1% and Czechia 4.4%; the ratio of those economically active in industry 37.5 vs. 37.7%; those commuting out of their residential areas 31.8 vs. 32.9%; inhabitants over the age of 60, 16.7 vs. 18.3% and ratio of recreational houses to total number in Ústí district 12.2% and in Czechia 13.8%.
2. In the Bílina area, seven settlements disappeared as a result of brown coal mining between the years 1964–1970. In the Petrovice area three settlements disappeared after the Second World War and three others are almost without any permanent population (under five inhabitants in 2001) – the result of undersettlement.

3. Symbol "+" represents an increase or stagnation of a given category in a reviewed period; symbol "–" represents decrease of a given category in the reviewed period. Category "+ + +"and "– – –" given under BTUs are theoretically impossible, because we compare territorially comparable BTUs.

References

Antrop, M. (2003). Continuity and change in landscape. In Ü. Mander & M. Antrop (Eds.), *Multifunctional landscapes. Vol. III – Continuity and change* (pp. 1–14). Southampton: WIT Press.

Antrop, M. (2005). Why landscapes of the past are important for the future. *Landscape Urban Planning, 70*(1), 21–34.

Balej, M. & Anděl, J. (2008). Land use changes and environmental stress accounting (case study from northwestern part of the Czech-German borderland). *Journal of Geography and Regional Planning, 1*(5), 97–109.

Bičík, I. (1997). Land use in the Czech Republic 1845–1948–1990. Methodology, interpretation, contexts. *AUC-Geographica, 32*(Supplementum), 255–263.

Bičík, I., Chromý, P., Jančák, V., Janů, & H. (Eds.). (2002) Land use/land cover changes in the period of globalization. *Proceedings of the IGU LUCC International Conference, Prague 2001.* Charles University, Prague.

Bičík, I., Götz, A., Jančák, V., Jeleček, L., Mejsnarová, L., & Štěpánek, V. (1996). Land-use/land cover changes in the Czech Republic 1845–1995. *Geografie-Sborník ČGS, 101*(2), 92–109.

Bičík, I., & Jančák, V. (2002). Development of agriculture and the Czech rural landscape. In K. Kobayashi (Ed.), *Changing regional structure and way of life in Central Europe: The case of Poland, the Czech Republic, the Slovak Republic and Hungary* (pp. 153–182). Gifu (Japan): Gifu University.

Douglas, I., Huggett, R., & Robinson, M. (1999). *Companion encyclopedia of geography. The environment and humankind.* New York: Routledge.

Feranec, J., & Ot'ahel, J. (1992). Land cover forms in Slovakia identified by application of colour infrared space photographs at the scale 1: 500 000. *Geografický časopis, 44*(2), 120–126.

Feranec, J., & Ot'ahel, J. (2000) *EEA Phare topic link on land cover (PTL/LC)*, Task 6 assessment of land cover and land use changes in the accession countries. Task 6. 1. Evaluation of methodology, Project Report.

Haberl, H., et al. (2002a). Land use change and socio-economic metabolism in Austria – Part I: Driving forces of land use change: 1950–1995. *Land Use Policy, 20*(1), 1–20.

Haberl, H., et al. (2002b). Land use change and socio-economic metabolism in Austria – Part II: Land use scenarios for 2020. *Land Use Policy, 20*(1), 21–40.

Haberl, H., Batterbury, S., & Moran, E. (2001). Using and shaping the land: A long perspective. *Land Use Policy, 18*(1), 1–8.

Haines-Young, R., & Potschin, M. (2003). Valuing and assessing of multifunctional landscapes: An approach based on the natural capital concept. In J. Brandt & H. Vejre (Eds.), *Multifunctional landscapes Vol. I – Theory, values and history* (pp. 181–192). Southampton: WIT Press.

Hampl, M. (1998). *Reality, society and geographical organisation. Finding an integral order.* Prague (in Czech): Charles University.

Himiyama, Y. (2002). Land use/cover change studies in the period of globalization. In I. Bičík, P. Chromý, V. Jančák, & H. Janů (Eds.), In *Land use/land cover changes in the period of globalization. Proceedings of the IGU-LUCC international conference Prague 2001* (pp. 5–14). Prague: Charles University.

Himiyama, Y., Mather, A., Bičík, I., & Milanova, E. V. (Eds.). (2005). *Land Use/Cover Changes in Selected Regions of the World* , IGU Study Group on LandUse/Cover Change, Hokkaido University of Education, Sapporo – Kita-ku.

Jensen, J. R. (2005). *Introductory digital image processing.* Upper Saddle River NJ: Prentice Hall.

Lambin, E. (2002). New global-scale datasets on land use/cover change. *LUCC Newsletter, 6*, 1–1.
Leppers, E. (2002). Millennium ecosystem assessment: Synthesis of knowledge of rapid land cover and land use change. *LUCC Newsletter, 6*, 2–3.
Lillesand, T. M., & Kiefer, R. W. (2002). *Remote sensing and image interpretation*. New York: John Wiley & Sons.
Lipský, Z. (1994). Landscape structure change of the Czech rural landscape. *Geografie Sborník ČGS, 99*(4), 248–260 (in Czech).
Lipský, Z. (1996). Land use changes and their environmental consequences in the Czech Republic. In R. Jongman (Ed.), *Ecological and landscape consequences of land use change in Europe* (pp. 350–360). Tillburg: ECNC.
Masek, F. (1948) *Plot cadastre*. Ministry of Finance, Prague (in Czech).
Meyer, W. B., & Turner, B. L., II (1996). The earth transformed: Trends, trajectories and patterns. In R. J. Johnston, P. J. Taylor, & M. Watts (Eds.), *Geographies of global change: Remapping the world in the late twentieth century* (pp. 302–317). Oxford: Blackwell.
Naveh, Z. (1991). Some remarks on recent developments in landscape ecology as a transdisciplinary ecological and geographical science. *Landscape Ecology, 5*(2), 65–73.
Naveh, Z. (2007). *Transdisciplinary challenges in landscape ecology and restoration ecology – An anthology*. Berlin, Heidelberg, New York: Springer.
Oršulák, T., Raška, P., & Suchevič, S. (2007). Reconstructive multidimensional geovisualisation of urban landscapes: Case study and prospects. *Historická Geografie (Historical Geography), 34*(1), 334–350 (in Czech).
Paine, D. P., & Kiser, J. D. (2003). *Aerial photography and image interpretation*. New York: John Wiley & Sons.
Palang, H., Printsmann, A., Konkoly Gyuró, E., Urbanc, M., Skowronek, E., & Woloszyn, W. (2006). The forgotten rural landscapes of central and Eastern Europe. *Landscape Ecology, 21*(3), 347–357.
Raška, P. (2006). Changes in a cultural landscape of the Doupov region – Specific case or the reflection of general political and social shifts. *Historická geografie (Historical geography), 33*(Supplementum), 162–174 (in Czech).
Skowronek, E., Krukowska, R., Swieca, A., & Tucki, A. (2005). The evolution of rural landscapes in mid-eastern Poland as exemplified by selected villages. *Landscape Urban Planning, 70*(1), 45–56.
Turner, B. W., II, et al. (1990). *The earth as transformed by human action: Global and regional changes in the biosphere over the past 300 years*. Cambridge UK: Cambridge University press.
Walsh, S. J., Evans, T. P., & Turner, B. L., II (2004). Population-environment interactions with an emphasis on land-use/land-cover dynamics and the role of technology. In S. D. Brunn, S. L. Cutter, & J. W. Harrington (Eds.), *Geography and technology* (pp. 491–519). Dordrecht: Kluwer.
Worster, D. (1985). *Natures' economy. A history of ecological ideas*. Cambridge UK: Cambridge University Press.
Žigrai, F. (1996). Integrative relevance of land use study in the regional geography. *Acta Facultatis Rerum Naturalium – Universita Comenianae-Geographica, 38*(Supplementum), 123–142.

Part IV
Changing Face of a Landscape: Identity and Perception

What does the landscape memory concept express? Is it possible to assess the memory or identity in an objective and exact way? Is the continuity in landscape development worthy of consideration for landscape management? How can single locations and objects affect regional identity? How is the land-cover change related to different functions and driving forces during history?

Chapter 8
Memory of a Landscape – A Constituent of Regional Identity and Planning?

Martin Balej, Pavel Raška, Jiří Anděl, and Alena Chvátalová

8.1 Landscape, Time and Man

In this chapter, we discuss Antrop's (2005) question: "Why are the landscapes of the past important for the future?" The definition of landscape in the Convention is clear and broad: "Landscape means an area, as perceived by people, whose character is the result of the action and interaction of natural and/or human factors" (Council of Europe, 2000). In this total human ecosystem (Naveh, 2003), a number of aspects of its constituent parts change with various spatiotemporal scales. Therefore, the landscape is changing and the landscape is also evolving. Research on landscapes of the past is important for future landscapes (sometimes referred to as "futurescapes"), and particularly for their planning and management.

But what is the bearer of information about the landscape's past? What makes a landscape in the eyes of its observer? On the basis of what information can we define its identity? The concept of *genius loci* is closely associated with the identity of each landscape and emphasizes its uniqueness (Antrop, 2000). A traditional landscape contains the comprehensive history of a place or region, which can still be read from its composition and structure (Antrop, 1997). Special places and monuments receive symbolic value and act as landmarks that allow orientation in space and time (Coeterier, 2002). One can say that the coherence of particular landscape properties defines identity. Changing the characteristics and coherence leads to a loss of identity or its change into a new one (cf. Soini, Palang, & Semm, 2006; Terkenli, 2006). Each landscape is unique in its own way because of its unique and incomparable composition and configuration and the character of the landscape elements that tell its story.

On one hand, there is the landscape, with information; on the other hand, there is the observer, who looks for, perceives and interprets this information. The observer does not approach the landscape as an indifferent objective viewer but as an active agent with subjective motives. Their interpretation of landscape information may differ from other observers' interpretations. The legibility of the information also

M. Balej (✉)
Department of Geography, Jan Evangelista Purkyne University in Ústí nad Labem, České mládeže 8, 400 96 Ústí nad Labem, Czech Republic
e-mail: martin.balej@ujep.cz

J. Anděl et al. (eds.), *Landscape Modelling*, Urban and Landscape Perspectives 8,
DOI 10.1007/978-90-481-3052-8_8,

plays a role, i.e. to what extent the information about the landscape's history is hidden in its component elements (and what their condition is). The landscape thus emerges as a historical space, as a narrative medium (Goldberg, Schwarz, & Porat, 2008), or, if you like, as the configuration of spatial narratives of time at historical sites (Azaryahu & Foote, 2008; Foote & Azaryahu, 2007).

We are interested in the spatial configuration of history – the way historical stories are arranged to be told in space to produce what we term the "spatial narratives" of history. Spatial narratives involve a complex configuration of geographic elements including buildings, markers, memorials, and inscriptions positioned with great care to provide a spatial storyline or to capture the key locational and chronological relations of a historical event. The proposition that a narrative is "anything that tells or presents a story, be it through text, picture, performance, or a combination of these [and] hence novels, plays, films, comic strips, etc. are all narratives" (Jahn, 2005) suggests that narratives are stories as presented through certain media. Paintings, sculptures and photographs cannot easily narrate an entire story, but can highlight key moments in the action that encapsulate, embody, symbolise or otherwise call to mind an entire plot (Lessing, 1962; Ryan, 2005).

Current landscape changes are seen as a threat, a negative evolution, because they are characterised by the loss of diversity, coherence and identity of the existing landscapes. The main difference between traditional and new landscapes resides in their dynamics, both in speed and scale, as well as the changing perceptions, values and behaviour of their users (Antrop, 2005). Nohl (2001) warns that the current landscape may lose its qualities. The other negative changes in the current landscape which he mentions include the lessening of regional identity or the loss of a rural structure. "Thus, the sense of place has gone, and landscape has lost its ability to tell specific and individual stories to the beholder" (Nohl, 2001). In the European context, research into landscape development, searching for the trajectories of landscape development and the determination of options for the scenarios of future landscapes often lead to implications similar to those mentioned above (e.g. Antrop, 2003; Balej, Anděl, Oršulák, & Raška, 2008; Bičík, Jeleček, & Štěpánek, 2001; Blaschke, 2003; Matless, 2008; Palang, 2005; Roca & Oliveira-Roca, 2007; Wagner & Gobster, 2007).

With regard to the above, we could define a landscape as a set of media which both individually and all together tell the story of the landscape. The current changes to the landscape and their characteristics, as often described, motivate us to try to shed more light on the issue of the narrative capacity of the landscape or the capacity of the landscape to inform the beholder about its history. We have attempted to define explicitly notions such as landscape memory, loss of landscape memory, and continuity of landscape development, which, in our opinion, are often used very vaguely.

The overall aim of our research is particularly connected to the answers to the following questions: What do the continuity and discontinuity of landscape development really mean? Is it the continuity of interwoven relations between man and a landscape? What are the indicators of developmental continuity of a landscape? Is it possible to find old lost villages, routes and other man-made landscape

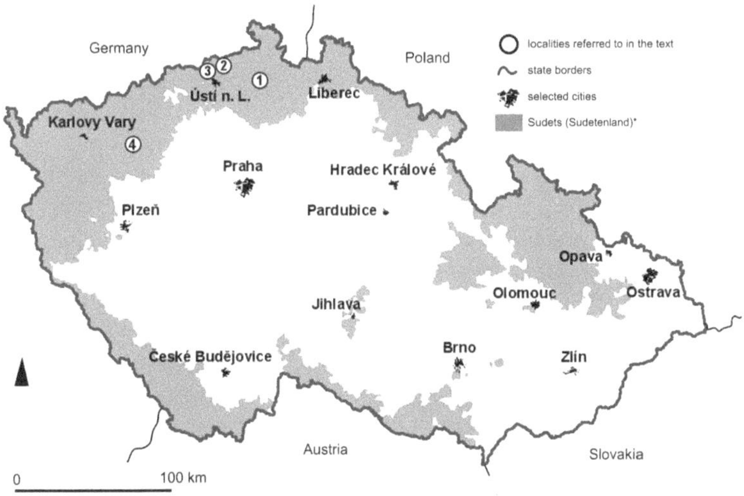

Fig. 8.1 The Sudetenland in the Czech Republic[1]

compositional elements to clarify the landscape memory? How useful is knowledge about the landscape memory?

The authors try to answer these questions on the micro-regional spatial scale, in case studies focused on several types of rural landscapes in the Czech–German borderland, the Sudetenland (Fig. 8.1). This area, with its unique history and "controversial" identity (Balej & Anděl, 2008), is studied at two essentially different stages of its development: (1) a period of relatively minor changes to the landscape and (2) a period of major changes to the landscape. By comparison with other areas in the Czech Republic, landscape changes were much more pronounced and extensive in the Sudetenland. Additionally, there was a marked concentration of the driving forces that determined this dynamic development.

The landscape in the period of relatively minor changes follows up on the Baroque landscape consolidated from the end of the Seventeenth century onwards, which displays a harmony between people and nature. Agricultural activity changes in the course of the Eighteenth century. Agricultural production intensifies, fallowing is replaced by fertilisation, and the rotation of crops becomes an established practice. As fallowing is abandoned, the acreage of fields increases over time by up to 50%. The deforestation process culminates, differentiation between meadows and pastures occurs (with regular two-cut management), land reclamation is applied (through canals with sluice gates) and the sacralisation of the landscape is completed (through small-scale sacral architecture, such as wayside columns, stations of the Cross, wayside crosses, crucifixes, chaplets, and pictures of saints). Efforts to reverse the shrinking of forests, particularly through reforesting thin deciduous forests with coniferous species, reach their zenith and result in pine and spruce monocultures.

Interest in landscape and nature in the Nineteenth century is of two types: on one hand, there is a romantic zeal for untouched wilderness, in which the countryside is a place to escape society, while on the other hand, townspeople show a growing interest in the countryside as a place for relaxation after their daily work activities. The first tourist guides emerge, and the idea of protecting remarkable, historic or untouched places gains ground with the development of industry and the natural sciences. The first tourist clubs are founded and people begin to build what is known today as tourist infrastructure. On the other hand, the processes of industrialisation and urbanisation, spreading from Saxon cities, are unfolding.

With the onset of Communism after World War II, the Czech landscape entered a breakthrough period – one of major landscape changes. The collectivisation and nationalisation of agricultural businesses and the introduction of socialist (i.e. collective) land ownership left permanent scars on the landscape that have still not healed. Following the Soviet model, individual fields were consolidated into a huge expanse of farmland. Balks were ploughed over and large land reclamation canals were built, accelerating the outflow of water from the natural landscape to trained, straight-line watercourses with paved-over concrete beds. Eighty percent of agricultural land was in the hands of farming cooperatives (called Standard Farming Cooperatives) in the period from 1955 to 1958. According to a 1958 central government directive on land consolidation, fields were to be optimised for the use of mechanised equipment, i.e. fields were required to have as few shape irregularities as possible. The directive also emphasised that crops were to be unified so that continuous blocks of fields were not disrupted by small forests, meadows or pastures.

This centrally controlled process had a much more profound impact on the Sudetenland. The reason was that some 3 million Germans were displaced from there after World War II. Some remote areas remained unsettled by the Czech inland population. Many settlements perished and almost one third of the agricultural land was left fallow. In the north-eastern Sudetenland, these negative trends were further exacerbated in the 1970s by an environmental crisis caused by heavily concentrated industrial enterprises. There were power plants burning brown coal extracted from large open-cast mines in the area, as well as intensive chemical production. The result was an overall deterioration of all the elements of the landscape (significant air pollution, degradation of soils, impaired vitality of the river systems and deterioration in the health of the forest stands).

8.2 The Concept of Landscape Memory

Before we can show the characteristics of landscape memory through actual examples and its interpretation for the purpose of landscape planning or for the development of options for landscape development scenarios, we need – as outlined in the previous section – to think more deeply about the concept of landscape memory and its former applications and we need to define it as a term.

The juxtaposition of the terms *landscape* and *memory* in landscape studies appears in various contexts and approaches. While the definition of the former is given great attention, also in response to the requirements laid down in legislation (Council of Europe, 2000), the use of the latter is rather intuitive and its unequivocal definition and application in landscape studies requires us to consider not only the landscape aspects but also the philosophical (Lothian, 1999) or environmental-psychological aspects (Ohta, 2001) of the issue.

As outlined in the previous section, specific research issues included in the concept of landscape memory have largely been concerned with the landscape as a narrative medium, where individual landscape features (sites, monuments) or the landscape's overall design are interpreted iconographically (cf. Cosgrove & Daniels, 1988), as symbols of events (e.g. Charlesworth, 1994) or as representations of power (Mitchell, 2002). Another topic to be dealt with in the perception of the landscape and its symbolic context is the relationship between landscape as a motif and art as a result (Andrews, 1999; Sandberg & Marsh, 2008). However, landscape memory, interpreted through the prism of the social sciences, cannot be understood only as the result of human activity having an effect on the landscape and our subsequent perception of the landscape. Rather, perception is a means of changing our thinking that can, in turn, change the landscape and impart to it a specific sense of place (Tuan, 1974; Urry, 1992; Allen, 1999). Similarly, in the classical work by Schama (1995), the whole cycle returns to the landscape or its natural foundations, which can become the starting point for human activity in the landscape. The ideal created through our perception of the landscape can then become the essence of our decision-making.

In contrast to this socially constructed approach, we can outline another approach, one that accentuates the landscape as a tangible product of historical processes resulting in unique physical transformations of the environment and new features in it. However, even these objects can have their abstract value for the sense of place. In this approach, the concept of *memory* is replaced by the more straightforward term *history* and the research methodology is based on geography, environmental studies and archaeology. The expansion of landscape historiography and landscape archaeology is primarily thanks to the British school. Classical works include Hoskins (1955); numerous contributions were later published in Landscape History (e.g. Hook, 2000). The natural environment and the ecological/environmental context in the study of historical societies were discussed by several authors (e.g. Butzer, 1964; Renfrew, 1983; Gojda, 2000).

There are certain differences between the two approaches. The former chiefly, though not solely, emphasises the current sociocultural interpretation of the past landscape, while the latter tries to reconstruct it. Despite these differences, however, both approaches focus primarily on the past landscape and for this reason they cannot be readily applied in landscape planning and the development of futurescapes (Antrop, 2005). Therefore, the concept of landscape memory needs to be approached in a different way, which hinges not on the anticipation of landscape content and its interpretation but on an analysis of the form – landscape components, i.e. what exists a priori before our interpretation (cf. Ohta, 2001).

First, we need to determine which perspective we intend to look at the term memory from. We can distinguish between memory as a quality (the capacity to remember), a process (the imprinting of specific features on one hand and recalling histories on the other) and an outcome (the outcomes of landscape transformation). It immediately becomes obvious that the logic of the terminological phrase "loss of memory", which appears in geographical papers, raises questions: "What kind of memory?" and "Can one of these be lost at all?"

While memory as a capacity and as a process emerges in a rather non-declaratory (non-conveyable) manner to the observer, memory as an outcome is declarative. Therefore, despite some of its imperfections, it appears to be the most reasonable perspective because an outcome is largely measurable. The structure of landscape memory includes five components: (a) genetic, (b) physical, (c) functional, (d) cultural and (e) informational, which will be explained and discussed below.

We will also explain that these components need to be understood as aspects of the evaluation of specific objects rather than attributing one component to one specific type of object.

The term *landscape memory* implicitly comprises time and for this reason we first need to determine the time scale and the related spatial scale, because different scales require different approaches and a different interpretation of the results. If we take memory as an outcome that manifests itself in a newly developed situation, the minimum time scale is one that allows a change which Antrop (2008) defines as "the difference in the state of an object, place or area between at least two different moments in time". However, the real scale that is necessary is going to be broader in order to allow the monitoring of development trajectories (the sequence of changes) in which we can trace both continuity and discontinuity. Both concepts have been amply discussed, particularly in the social sciences (e.g. Lowie, 1987). In the 1950s, Hoskins (1955) saw landscape development as a continuous sequence of changes without a standstill period. An uninterrupted sequence of changes could then be seen as an absence of major breakthrough moments. The disadvantages of the concept of "discontinuity" and efforts to identify development milestones were discussed in historiography by Le Goff (2003), who noted the relativity of the time periods so determined and applied to various territories or looked upon from various perspectives. However, the spatial scale assumed here is the landscape (according to Hobbs, 1997; Council of Europe, 2000) and the purpose is not to delimit time periods but rather to look for changes which were significant enough to disrupt the continuous development of the landscape (Antrop, 2005). The aim is to identify the consequences of these discontinuities in the above-mentioned components of the memory of a specific landscape.

In the following sections, we will identify the individual components of landscape memory and provide examples that have been analysed primarily by means of historic geographical methods (old maps and photos, historical statistics, etc.) and landscape-ecological methods (field mapping of anthropogenic transformation, geomorphic and botanical surveys, etc.).

8.3 Components of Landscape Memory: Between Linkages and Contradictions

Following a structural analysis of landscape objects and the aspects through which they can be examined, the components of landscape memory can be determined as follows. The genetic component embodies the natural environment (background) of human activity. The physical component is the sum of the real, existing features of seminatural or anthropogenic origin that occur in the landscape. The functional component embodies the way in which the landscape is used, not only in the traditional sense of land-use categories but as a whole array of actual modes of use of landscape features. The cultural component denotes the cultural historical substance (value) of landscape memory features and the information component denotes the potential to convey information about the landscape's history.

None of the above-mentioned components is related to only one type of feature in the landscape. For instance, natural monuments are not only an expression of the natural environment (genetic component) but also have their physical, cultural and other components. Similarly, a historical building is a physical object but also has a functional component (e.g. a residence or museum) or information component (e.g. iconographic or educational value).

For this reason, the components of landscape memory represent aspects (or values) that are characteristic of each object, while the components of various features can assume various meanings. Additionally, the components can be in relative harmony or, on the contrary, in sharp contradiction with each other, which is documented by the examples provided below.

The characteristic elements of a rural landscape include line objects, whose primary purpose was usually to divide up land in historical times. The best-known example would probably be the hedgerows in England (Oreszczyn & Lane, 2000) but other examples are known from other European countries as well (e.g. Sitzia, 2007; Sklenicka et al., 2009). In the Sudetenland, these field boundaries have the character of dry stone walls which, in the absence of human intervention, were gradually covered by abundant communities of tree and herbal species. Examples of such line objects, commonly termed agricultural levees, can be seen in Fig. 8.2, shown as Locality 1 in Fig. 8.1.

As with hedgerows, the function of agricultural levees has shifted from land delimitation to an ecological function, where the levees serve as corridors for species (Forman & Baudry, 1984; Sitzia, 2007; Roy & de Blois, 2008). However, Oreszczyn and Lane (2000) also point out that such a restrictive understanding does not take into account the wealth of information that these line elements in the landscape can convey. This is why they also study hedgerows as part of national identity, a bond with the past, a contribution to the sense of place etc.

Similarly, agricultural levees complete the overall landscape design by incorporating, apart from the physical component of landscape memory (i.e. their existence), a functional (ecological) component, a cultural component (specific

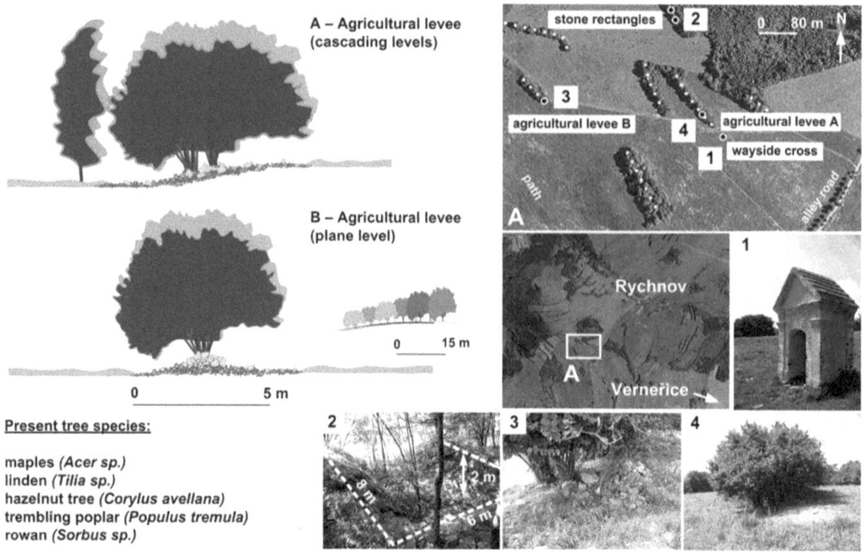

Fig. 8.2 Hedgerows as an example of a landscape element in which we can identify all the five components of landscape memory

fragmentation of the landscape, which influences regional cultural identity) or an information component (information about former ways of agricultural management). The only transformed component is the functional component but this is more a shift in weight rather than a replacement because both the delimitation function and the ecological function can be performed at the same time. From this point of view, all the components are in a clear relation to each other and from a long-term perspective the development of agricultural levees as landscape components is not discontinuous.

A similar example would be that of cemeteries, which represent unique multifunctional landscape features. The primary implicit function of cemeteries is to embody memory (of life and landscape) and they are simultaneously an architectural testimony to the cultural history and religiousness of an area (Hupková & Havlíček, 2008), a source of information about the character of historical and more recent societies and communities (Hristova, 2006; Miller & Rivera, 2006; Rugg, 2006) and, last but not least, a feature performing ecological functions. The last-mentioned functions are made possible by the fact that cemeteries form an area that is inwardly heterogeneous but outwardly sharply delimited (even more so in an urban landscape), with relatively slow development. Similarly as in the study by Westcoat, Brand, & Mir, (1991), cemeteries can also be interpreted not only as surface systems interconnecting ecological and sociocultural values but also as real three-dimensional systems reaching both above and below the ground – it is in this intersection that their essence lies. On this basis, we can observe the physical, functional, cultural and informational components of landscape memory.

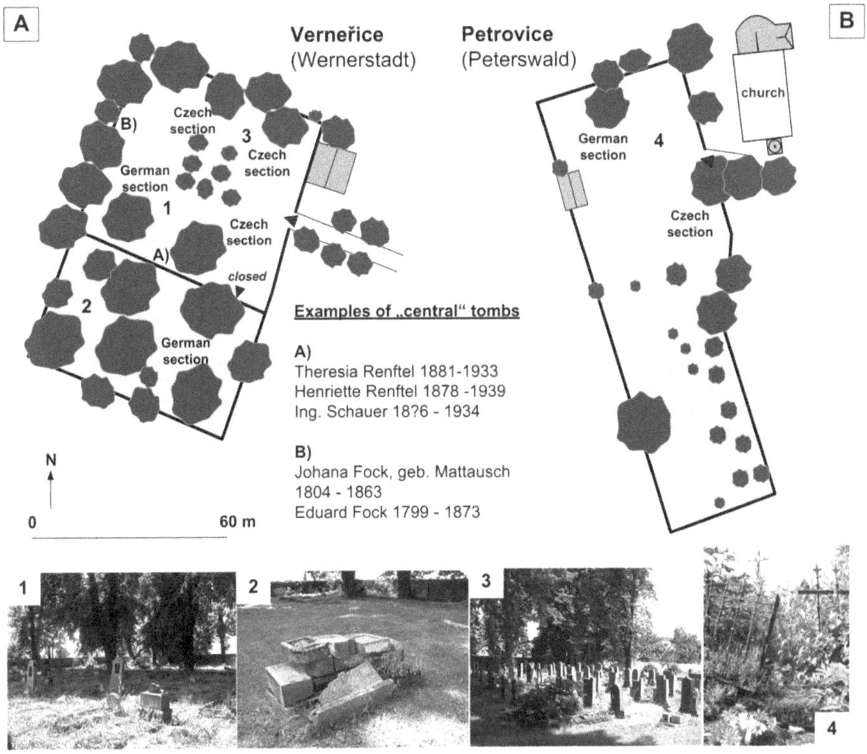

Fig. 8.3 A cemetery as an example of a landscape element in which all the five components of landscape memory can be identified

Figure 8.3 shows the layout of the cemetery in the villages of Verneřice and Petrovice (Localities 1 and 2 on the map in Fig. 8.1). The village of Verneřice is a typical settlement representing the sociocultural and landscape development of rural areas in the Sudetenland. The oldest tombstones preserved in the cemetery located on the outskirts of the village date back to the second half of the Nineteenth century. Thanks to this, the cemetery provides a great opportunity to monitor the impact of political events (German settlement, World War II, displacement of the German population etc.) on the development of individual features in the landscape. The cemetery is divided into two sections, separated from each other by a wall. The smaller section is no longer used and only contains the tombstones of the local German population of the pre-war period. The entrance to this section is permanently closed, the tombstones and the walls have fallen into disrepair and the vegetation there is subject to natural succession, interrupted only by occasional mowing. The other, larger section is still used and serves the current population of the village. Despite this, it contains not only the tombstones of deceased Czech villagers but also parts with German tombstones as well, some standing, others dilapidated or fallen. In terms of landscape memory, this can be interpreted as follows. The physical component, which clearly acts as the symbol of discontinuity, is the wall with the permanently closed entrance to the smaller section of the cemetery that is no longer in use. The components of memory are therefore rather disproportionate. Funerals only take place in the larger section of the cemetery (the primary function of burying), while the cultural and informational components have been preserved in the other section as well.

For instance, family names such as Renftel or Mattausch on the oldest tombstones (see Fig. 8.3) are mentioned in historical sources about the village as early as the mid-Seventeenth century. However, there is a question as to how long this will remain the case, given the gradual deterioration of the cemetery. Another example of indirect information about the past that has been preserved is the location of the cemetery itself. From there, we can see traces of roads marked on older (military) maps that lead to a site of pilgrimage on the hilltop of Gottesberg with a church, which has now vanished. The neighbouring village of Petrovice offers an interesting comparison. Petrovice is a border settlement between the Czech lands and Germany. A spatial analysis and field survey have shown the significance of the impact of the distance of the community from the border. Unlike the village of Verneřice, the former German inhabitants or their descendants have been visiting and maintaining some of the tombs since the 1990s. As a whole, the spatial structure of the cemetery in Petrovice is much more complex than that of the one in Verneřice and it does not contain strictly delimited national sections.

While the first example showed functioning relations between the individual components and the second a partial connection and partial contradiction between them, there are numerous localities in the Sudetenland where the components of landscape memory are in sharp contradiction with each other. Localities 1, 2 and 3 (Fig. 8.1) feature churches for which a lot of effort has been put into external maintenance. However, the furnishings inside are non-existent and in one particular case the roof over the main nave of the church could not be preserved either. The physical and cultural component (although disrupted by the damage to the buildings) is thus in complete contradiction with the functional component of the ceremonial use of the buildings. Locality 4 (Fig. 8.1) and perhaps other localities in the area (Raška, 2006) have probably gone through the most complicated development. The German inhabitants were originally the dominant majority population here. After their post-war displacement, the area was repopulated by inhabitants from the central part of Bohemia or Eastern Europe. However, a military zone was created here in 1953 and the area was depopulated again, with the exception of a few villages on its periphery. Buildings disintegrated over time or were demolished, of the original road system only those routes used for the new purposes were maintained, and the mosaic of agricultural areas became covered up by vegetative succession. Despite the ecological value of some of the succession communities of species in the locality (as well as in the wider area), the components of landscape memory are in contradiction with each other here and some of them cannot be traced any longer.

8.4 Relevance of the Concept in Landscape Planning

The search for all the five components of landscape memory brings us a vital understanding about landscape development, about the landscape of the past. Researchers can read the narratives of landscape history. The result of their research is not and cannot be a simplified statement about the loss of landscape memory. The narratives remain in the landscape. Some may perish, others may emerge. Only the story told by each of them individually and all of them together is changing. A complete loss of landscape identity cannot be the conclusion either, because the landscape may acquire a new identity. The purpose of the scholarly examination of, and search for, the components of landscape memory lies in the interpretation of landscape history. How was the story of the past landscape reflected in its functioning? What was the impact on biodiversity, ecohydrological functions, stability, soil fertility, and the quality of the natural components of the environment (air, ecosystems etc.)? How did it influence the social environment? Did it provide enough job opportunities and, as a whole, allow well-being and healthy living? Did it create a harmonious interconnection between the social and the ecological subsystems so that we can label them sustainable (cf. Balej & Anděl, 2008)?

Interpretations along these lines provide many desirable answers to landscape management and landscape planning. Scientists need to listen in order to understand what policy-makers care the most about. Understanding needs and beliefs will allow scientists to design their research so that it is truly relevant and salient to policy-makers. It is particularly vital to include the viewpoints of the land users themselves throughout the process (Reid et al., 2006). Without the inclusion of interpretations based on the knowledge of the land users themselves it is not possible to justify consistently any recommendation regarding the optimum landscape development scenario or plan.

Policy research that aims to be useful to policy-makers starts with a clear definition of a policy research problem, including an assessment of policy objectives and the impact of existing policies, the identification of relevant policy instruments, and establishing working relationships with the policy-makers who have influence over those policy instruments. Land use scientists need to work closely with policy-makers and land users to identify – and in many cases develop, test, and validate – workable policy levers that effectively influence the rate and patterns of land-use change (Tomich et al., 2004).

As examples of landscape narratives, agricultural levees, road networks, greenways, hedgerows, sacral elements, and cemeteries, as well as the urban planning and architectural concepts of settlements, allow us to monitor and assess the various manifestations of landscape memory components and to examine their conformity or contradiction with each other. A harmonious landscape development or, better still, the continuity of landscape development can be defined as a mutual reflection of the historical context and its manifestation, i.e. as a consonance between all the five components (aspects) of landscape memory in the individual landscape elements and in the landscape as a whole. Specific driving forces (proximate and underlying causes, according to Mather, 2006) may put these aspects of landscape elements into contradiction, thereby diminishing their mutual harmony. If this contradiction between the landscape elements exists in the landscape to a great extent (i.e. in terms of intensity or area) or if it affects a significant number of landscape elements or if it affects the key elements (the dominant landscape features), we can say that the landscape is beginning to tell another story.

The notion of the loss of landscape memory means the absence of certain type(s) of landscape element or the loss of their ability to bear some of the original five components of landscape memory. However, the memory of the landscape as such is not lost. The narrative capacity of the landscape continues to exist but the story that the landscape has told about itself so far is beginning to blur or a new story is emerging. In fact, these landscape elements (or their traces in the landscape) often remain physically in place and only some of the other four components of landscape memory perish or are transformed into a contradictory form. Landscape memory is similar to human memory. If we do not recollect, recall, or re-live our previous experiences, they are left abandoned in our subconscious mind until another intense experience brings them back. Landscape memory as a capacity does not perish. But the stories the landscape tells may be transferred into some kind of "landscape subconsciousness". However, they are not lost forever; they can be retrieved and recalled from there.

The essence of optimum landscape planning (Antrop, 2005) should be in the reflection of the context of historical development. This is not a call for us to transpose a historical condition of the landscape to the present day. It is an assessment of the results of interpretations of the landscape of the past in terms of sustainability and multifunctionality (Potschin & Haines-Young, 2003). Both the coherence of individual landscape narratives between each other and the narratives themselves in time are the co-bearers of landscape identity. Landscape identity is created by the observer (the reader of landscape stories) with their approach, through the perception of the landscape on the basis of both individual (ontological) and collective (historical) experience. Disparities in these experiences (e.g. new settlers in a formerly depopulated area) weaken regional identity (Bičík et al., 2001; Palang et al., 2005; Henige, 2007). The co-bearers of this identity were forced out of the landscape. A connection between the interpretations of landscape history and regional identity may be, for instance, a landscape character that can be articulated as a visual quality or landscape aesthetics (Nohl, 2001; Lothian, 1999).

Acknowledgements The research presented was supported by the research project of the Ministry of Labour and Social Affairs of the Czech Republic (No. 1J 008/04-DP1) and project *Czech Borderland after Schengen: a Distinct, Oscillating and/or Transit Area?* (No. IAA311230901) supported by the Grant Agency of the Czech Republic.

Notes

1. Sudets (Sudetenland) – from now on to be understood as the area with a pre-war prevalence of German population situated in the border or near-to-border areas of the country.

References

Allen, T. (1999). The management of the rural landscape: A sense of place. In J. Grenville (Ed.), *Managing the historical rural landscape* (pp. 163–162). London: Routledge.

Andrews, M. (1999). *Landscape and western art*. Oxford: Oxford University Press.

Antrop, M. (1997). The concept of traditional landscapes as a base for landscape evaluation and planning. The example of Flanders region. *Landscape and Urban Planning, 38*(1–2), 105–117.

Antrop, M. (2000). Where are the genii loci?. In B. Pedroli (Ed.), *Landscape – our home, lebensraum landschaft. Essays on the culture of the European landscape as a task* (pp. 29–34). Zeist, Freies Geistesleben, Stuttgart: Indigo.

Antrop, M. (2003). Continuity and change in landscape. In Ü. Mander & M. Antrop (Eds.), *Multifunctional landscapes. Vol. III – Continuity and change* (pp. 1–14). Southampton: WIT Press.

Antrop, M. (2005). Why landscapes of the past are important for the future?. *Landscape and Urban Planning, 70*(1–2), 21–34.

Antrop, M. (2008). Landscapes at risk: About change in the European landscape. In P. Dostál (Ed.), *Evolution of geographical systems and risk processes in the global context* (pp. 57–80). Prague: Charles University.

Azaryahu, M., & Foote, K. E. (2008). Historical space as narrative medium: On the configuration of spatial narratives of time at historical sites. *GeoJournal, 73*(3), 179–194.

Balej, M., & Anděl, J. (2008). Land use changes and environmental stress accounting (case study from northwestern part of the Czech-German borderland). *Journal of Geography and Regional Planning, 1*(5), 97–109.

Balej, M., Anděl, J., Oršulák, T., & Raška, P. (2008). Development of environmental stress in the northwestern part of Czechia: New approaches and methods. *Geografie – Sborník ČGS, 113*(3), 320–336.

Bičík, I., Jeleček, L., & Štěpánek, V. (2001). Land-use changes and their social driving forces in Czechia in the 19th and 20th centuries. *Land Use Policy, 18*(1), 65–73.

Blaschke, T. (2003). Continuity, complexity, and change: A hierarchical geoinformation-based approach to explore patterns of change in cultural landscape in Germany. In Ü. Mander & M. Antrop (Eds.), *Multifunctional landscapes. Vol. III – Continuity and change* (pp. 33–54). Southampton: WIT Press.

Butzer, K. W. (1964). *Environment and archaeology. An ecological approach to prehistory*. Chicago: Aldine Press.

Charlesworth, A. (1994). Contesting places of memory: The case of Auschwitz. *Environment & Planning D: Society & Space, 12*(5), 579–593.

Coeterier, J. F. (2002). Lay people's evaluation of historic sites. *Landscape and Urban Planning, 59*(2), 125–127.

Cosgrove D., Daniels S. (Eds.). (1988). *The iconography of landscape*. Cambridge: Cambridge University Press.

Council of Europe (2000) *The European Landscape Convention*, Strasbourg.

Foote, K., & Azaryahu, M. (2007). Toward a geography of memory: Geographical dimensions of public memory and commemoration. *Journal of Political and Military Sociology, 35*(1), 125–144.

Forman, R. T. T., & Baudry, J. (1984). Hedgerows and hedgerow networks in landscape ecology. *Environmental Management, 8*(6), 495–510.

Gojda, M. (2000). *The archaeology of landscape*. Praha (in Czech): Academia.

Goldberg, T., Schwarz, B. B., & Porat, D. (2008). Living and dormant collective memories as contexts of history learning. *Learning and Instruction, 18*(3), 223–237.

Henige, D. (2007). 'This is the place': Putting the past on the map. *Journal of Historical Geography, 33*(2), 237–253.

Hobbs, R. J. (1997). Future landscapes and the future of landscape ecology. *Landscape and Urban Planning, 37*(1), 1–9.

Hook D. (Ed.). (2000). Landscape: The richest historical record. *Landscape History – Journal of the Society for Landscape Studies*, Supplementary Series 1.
Hoskins, W. G. (1955). *The making of the English landscape*. London: Hodder.
Hristova, S. (2006). Images of social inequality in a Bulgarian Transition city – The market and the cemetery. *Research in Urban Sociology, 8*, 247–264.
Hupková, M., & Havlíček, T. (2008). Religious landscape in Czechia: New structures and trends. *Geografie-Sborník ČGS, 113*(3), 302–319.
Jahn, M. (2005). *Narratology: A guide to the theory of narrative*. Cologne: University of Cologne.
Le Goff, J. (2003). *À la recherche du Moyen Âge*. Paris: Seuil.
Lessing, G. (1962). *Laocoon. An essay on the limits of painting and poetry*. Indianapolis: The Boss-Merrill Company.
Lothian, A. (1999). Landscape and the philosophy of aesthetics: Is landscape quality inherent in the landscape or in the eye of the beholder?. *Landscape and Urban Planning, 44*(4), 177–198.
Lowie, R. H. (1987). The determinants of culture. In H. Applebaum (Ed.), *Perspectives in cultural anthropology* (pp. 85–99). New York: SUNY Press.
Mather, A. S. (2006). Proximate causes. In H. J. Geist (Ed.), *Our earth's changing land: An encyclopedia of land use and land cover change, Vol. 2 (L-Z)* (pp. 490–495). Westport, London: Greenwood Press.
Matless, D. (2008). Properties of ancient landscape: The present prehistoric in twentieth-century Breckland. *Journal of Historical Geography, 34*(1), 68–93.
Miller, D. S., & Rivera, J. D. (2006). Hallowed ground, place, and culture. The cemetery and the creation of place. *Space and Culture, 9*(4), 334–350.
Mitchell W. J. T. (Ed.). (2002). *Landscape and power*. Chicago: University of Chicago Press.
Naveh, Z. (2003). The importance of multifunctional self-organising biosphere landscapes for the future of our total human ecosystem – A new paradigm for transdisciplinary landscape ecology. In J. Brandt & H. Vejre (Eds.), *Multifunctional landscapes Vol. I – Theory, values and history* (pp. 33–62). Southampton: WIT Press.
Nohl, W. (2001). Sustainable landscape use and aesthetic perception-preliminary reflections on future landscape aesthetics. *Landscape and Urban Planning, 54*(1), 223–237.
Ohta, H. (2001). A phenomenological approach to natural landscape cognition. *Journal of Environmental Psychology, 21*(4), 387–403.
Oreszczyn, S., & Lane, A. (2000). The meaning of hedgerows in the English landscape: Different stakeholder perspectives and the implications for future hedge management. *Journal of Environmental Management, 60*(1), 101–118.
Palang, H., et al. (2005). The forgotten rural landscapes of central and Eastern Europe. *Landscape Ecology, 20*(3), 347–357.
Potschin, M. B., & Haines-Young, R. H. (2003). Improving the quality of environmental assessment using the concept of natural capital: A case study from southern Germany. *Landscape and Urban Planning, 63*(2), 93–108.
Raška, P. (2006). Changes in a cultural landscape of the Doupov region – Specific case or the reflection of general political and social shifts. *Historická geografie (Historical geography), 33*(Supplementum), 162–174 (in Czech).
Reid, R. S., Tomich, T. P., Xu, J., Geist, H., Mather, A., DeFries, R. S., et al. (2006). Linking land change science and policy: Current lessons and future integration. In E. F. Lambin & H. J. Geist (Eds.), *Land-use and land-cover change* (pp. 157–171). Local Processes and Global Impacts. Germany: Springer.
Renfrew, C. (1983). Geography, archaeology and environment – I. Archaeology. *The Geographical Journal, 149*(3), 316–333.
Roca, Z., & Oliveira-Roca, M. D. N. (2007). Affirmation of territorial identity: A development policy issue. *Land Use Policy, 24*(2), 434–442.
Roy, V., & de Blois, S. (2008). Evaluating hedgerows corridors for the conservation of native forest herb diversity. *Biological conservation, 141*(1), 298–307.
Rugg, J. (2006). Lawn cemeteries: The emergence of a new landscape of death. *Urban History, 33*(2), 213–233.

Ryan, M. (2005). Media and narrative. In D. Herman, M. Jahn, & M. Ryan (Eds.), *Routledge encyclopedia of narrative theory* (pp. 288–292). London: Routledge.

Sandberg, L. A., & Marsh, J. S. (2008). Literary landscapes – geography and literature. *Canadian Geographer, 32*(3), 266–276.

Schama, S. (1995). *Landscape and memory*. London: HarperCollins.

Sitzia, T. (2007). Hedgerows as corridors for woodland plants: A test on the Po Plain, Northern Italy. *Plant Ecology, 188*(2), 235–252.

Sklenicka, P., Molnarova, K., Brabec, E., Kumble, P., Pittnerova, B., Pixova, K., et al. (2009). Remnants of medieval field patterns in the Czech Republic: Analysis of driving forces behind their disappearance with special attention to the role of hedgerows. *Agriculture, Ecosystems and Environment, 129*(4), 465–473.

Soini, K., Palang, H., & Semm, K. (2006). From places to non-places? Landscape and sense of place in the finnish and estonian countrysides. In T. S. Terkenli & A. -M. d'Hauteserre (Eds.), *Landscapes of a new cultural economy of space* (pp. 117–148). Berlin, Heidelberg, New York: Springer.

Terkenli, T. S. (2006). Landscapes of a new cultural economy of space: An introduction. In T. S. Terkenli & A. -M. d'Hauteserre (Eds.), *Landscapes of a new cultural economy of space* (pp. 1–18). Dordrecht: Springer.

Tomich, T. C., Thomas, D. E., & van Noordwijk, M. (2004). Environmental services and land use change in Southeast Asia: From recognition to regulation or reward? *Agriculture, Ecosystems and Environment, 104*(1), 229–244.

Tuan, Y. -F. (1974). *Topophilia: A study of environmental perception, attitudes, and values*. Englewood Cliffs (NJ): Prentice-Hall.

Urry, J. (1992). The tourist gaze and the "Environment". *Theory, Culture & Society, 9*(1), 1–26.

Wagner, M. M., & Gobster, P. H. (2007). Interpreting landscape change: Measured biophysical change and surrounding social context. *Landscape and Urban Planning, 81*(1–2), 67–80.

Westcoat, J. L., Jr, Brand, M., & Mir, N. (1991). Gardens, roads and legendary tunnels: The underground memory of Mughal Lahore. *Journal of Historical Geography, 17*(1), 1–17.

Chapter 9
Landscape Change in the Seewinkel: Comparisons Among Centuries

Martin A. Prinz, Thomas Wrbka, and Karl Reiter

9.1 Existing Research and Specific Research Question

Landscape pattern displays the interaction between natural and cultural forces (Baudry, 1989; Naveh, 1995; Bürgi, Hersperger, & Schneeberger, 2004; Haber, 2004). In times of changing land-use systems (intensification and unification), climate change and biodiversity loss it is necessary and helpful to know how landscapes were built up in former times to develop potential scenarios for the future (Antrop, 2004, 2005). This change is, as is typical for most landscapes, a slow ongoing process which can only be seen by comparing data from different centuries. Various studies have dealt with this issue, but most of them tried to picture changes over decades (e.g. Jenerette & Wu, 2001; Nelson, Soranno, & Qi, 2002; Crews-Meyer, 2004) not over centuries (e.g. White & Mladenoff, 1994; Tasser, Teutsch, Noggler, & Tappenier, 2007). Also the Seewinkel, an area east of Lake Neusiedl has been partly analysed (Kohler, Rauer, & Wendelin, 1994).

Land-use change has been shown as an important driver for biological impoverishment, especially in European agricultural landscapes (Abensperg-Traun, Wrbka, Bieringer, & Hobbs, 2004). Therefore, modern conservation strategies need a state of the art documentation of not only recent but also historical trajectories of landscape change.

On the basis of the interpretation of maps, this kind of analysis was only possible because of the existing maps of the Military Surveys (MS) from the eighteenth and nineteenth century. Like other old maps (Cajthaml, 2007a; Krejčí & Cajthaml, 2007) these data have already been used in many projects for answering various ecological and historical questions concerning land use and land cover throughout the area of the former Austrian-Hungarian Monarchy (e.g. Kozak, 2003; Dömötörfy, Reeder, & Pomogyi, 2003; Kovács, Zámolyi, Székely, & Papp, 2008; Timár & Pišút, 2008; Timár et al., 2008).

M.A. Prinz (✉)
Department of Conservation Biology, Vegetation & Landscape Ecology,
University of Vienna, Rennweg 14, A-1030 Vienna, Austria
e-mail: martin.prinz@univie.ac.at

J. Anděl et al. (eds.), *Landscape Modelling*, Urban and Landscape Perspectives 8,
DOI 10.1007/978-90-481-3052-8_9, © Springer Science+Business Media B.V. 2010

The aim of this study was to point out the change of land use and land cover in the Seewinkel during the last 250 years. Further aims were to highlight changes in agricultural system and to illustrate processes of intensification and industrialisation.

9.2 Regional Settings

The study area, the so-called "Seewinkel" is a small part of the federal state Burgenland (Austria) situated east of the eastern shore of Lake Neusiedl. In the south and east it builds up the border to Hungary and enters the so-called "Waasen" (Hungarian: Hanság), a former large-scale fen area. The area can be mainly assigned to four types of cultural landscapes: (1) Extra-alpine basins and valley floors with dominant crop farming; (2) Pannonian arable – viticulture complex; (3) Flatlands and soft slopes with dominant viticulture and (4) Extended extra-alpine xeric grassland and pasture landscapes (Wrbka et al., 2002). The partly swampy area contains more than 40 temporary shallow salty lakes and borders to the most western salt lake in Europe (Berger, Fally, & Lunzer, 1992). It is the absolute lowest (on average only 117 m above sea level) and one of the driest (annual precipitation <600 mm/a – ZAMG 2002) parts of Austria and has been colonised since the Neolithic period (Löffler, 1982). Gravel dominates the bedrock (Haeusler, 2006; Löffler, 1982). The main soil types are different kinds of tschernosem and salt-influenced soil types like solonetz or solontschak (Nestroy, 2005; Löffler, 1982). In former times the grassy steppe around these lakes allowed an extensive stock-breeding (especially cattle and horses). Today's use of these particular areas is mostly dedicated to small-scale agriculture or tourism.

Generally this former wooded steppe has been transformed through millennia of agricultural use into a more or less steppic area without natural forests and was especially modified during the last 250 years of cultivation.

Particularly valuable for nature conservation are (1) the seasonally flooded area in front of the lake, (2) the reed belt, (3) the small shallow lakes and (4) different types of extensively used meadows. Most of the shallow lakes dry out in summer to salty swamps or desert-like areas covered by a salt crust. The physico-chemical secret of these rain-determined ecosystems lies in the sealing function of the sodium salt in the substrate. This system only works when the substrate is wet from a high groundwater level. When the groundwater level declines, the sodium salt substrate dries out and the water impermeable horizon becomes crumbly and permeable.

The protected landscape around Lake Neusiedl is the largest Ramsar site in Austria with 44.229 ha (Umweltbundesamt, 2004). In 1993 parts of the Seewinkel and the nature protection areas of Lake Neusiedl were declared together with the Hungarian part Fertö-Hanság (National Park since 1991) as a trans-boundary National Park and in 2001 as World Natural Heritage (Nationalpark Neusiedlersee-Seewinkel, 2008).

9.3 Material and Methods

9.3.1 Historical Maps

Digitised historical maps were the basis for the representation of the landscape change (Křováková & Brůna, 2007; Timár & Pišút, 2008; Timár et al., 2008; Mikšovský & Zimová, 2007). On the one hand sheets of the 1st Military Survey were used, which was accomplished between 1763 and 1785 in the Austro-Hungarian monarchy. Sheets from this particular part of the monarchy were plotted in 1785 (Kretschmer & Riedl, 2008). Especially concerning their colour these sheets were high quality but to some extent there is a remarkable difference in terms of the content (meadow turns to arable land at the sheet border, tracks stops abruptly, etc.) of the individual sheets. On the other hand there were sheets available from the 2nd MS (accomplished 1821–1869, particularly plotted in 1845). The quality of these sheets varies strongly. Especially the southern most sheets are low grade concerning colour. The central sheets have almost the same quality as the sheets from the 1780s. The original northern-most sheets got lost and were obviously only available as bad brown-coloured copies of the original. Content differences between the sheets were not as serious and rather rare (for instance draw wells are missing completely in the western sheets). Because of the fact, that it is not possible to rectify this kind of data accurately into present-day coordinate systems (Timar et al., 2007; Cajthaml, 2007b) the maps were rectified manually via ground control points. To increase the precision of rectification an already existing high-resolution laser-scan (LIDAR) was used to build a digital terrain model (DTM) with a horizontal resolution of 17 and 4 cm accuracy in height (Bitenc, 2007). Military surveys on the whole concentrate very strongly on militarily relevant land marks. In this special case a lot of the plotted small observation knolls can be accurately linked to uprisings in the DTM.

9.3.2 Current Land Cover Data

Current land cover data were derived from the official digital cadastral map (DCM, which is constantly renewed) and SINUS land-cover data (Peterseil, et al. 2004). Because of poor resolution and too common categories within these two datasets additional data from various practical courses and field surveys were included especially for displaying actual grassland areas properly.

9.3.3 Determination of Investigation Area

The investigation area has been defined as the extension area of the used DTM. This area is restricted to a broad strip along the eastern shore of Lake Neusiedl up to the Hanság on the Hungarian border. Figure 9.1 displays the extension of the investigation area. Altogether the analysed area covers 317.18 km^2. Different kinds of overlapping nature protection areas can be found (see Table 9.1).

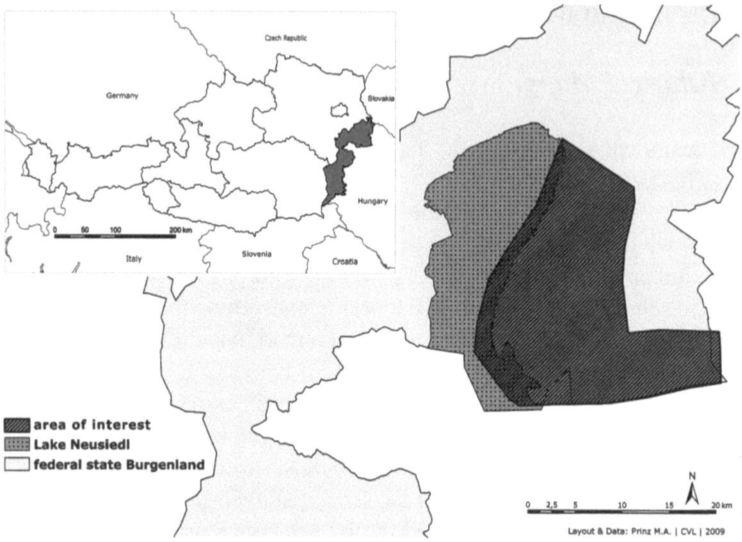

Fig. 9.1 Investigation area

Table 9.1 Protection areas and proportion of investigation area

Nature protection category	Coverage (%)
Biosphere reserve	12.57
Landscape-protected area	52.17
National Park	14.65
Nature conservation area outside National Park	0.06
Areas according to European Bird Directive	52.17
Areas according to Natura 2000 (FFH) – Directive	0.00

9.3.4 Digitalisation and Interpretation

Minimal mapping unit represents the smallest visible structure. Therefore streets (tracks) and small drains were also digitised. To make this process easier, rivers (only visible in sheets from the 1780s) and traffic routes were digitised as line first and finally buffered to a 3 m extent (real width then 6 m irrespective of schematic drawn width). When digitising maps from the 1840s the same procedure was adapted for drawn in dams (also 6 m width). Point-shaped landscape elements (military knolls for border observations or cairns of the Iron Age and Roman times in the first instance) were digitised as points and then buffered to an extent of 40 m (equivalent to approx. 5.000 m^2). Together with the plane elements these two shapes were finally merged geographically together. All generated midget-areas smaller than 100 m^2 were assigned to the largest adjacent polygon.

Because of their universally applicable attributes the basis for interpretation was the attributes of the CORINE land-cover classification (Bossard, Feranec, & Otahel,

2000). Signatures used in the sheets are very comprehensive, especially from the 2nd Military Survey. Hungarian signatures have been translated into German to make interpretation easier.

9.3.5 Analysis

To make a comparison of old data with current land cover data possible several utilisation classes were combined to utilisation blocks (see Fig. 9.2). Areas with other usage contain (1) mineral extraction sites, (2) beaches, dunes and sandplains, (3) other sparely vegetated areas, (4) bosks and groves, (5) forests and (6) different kinds of gardens.

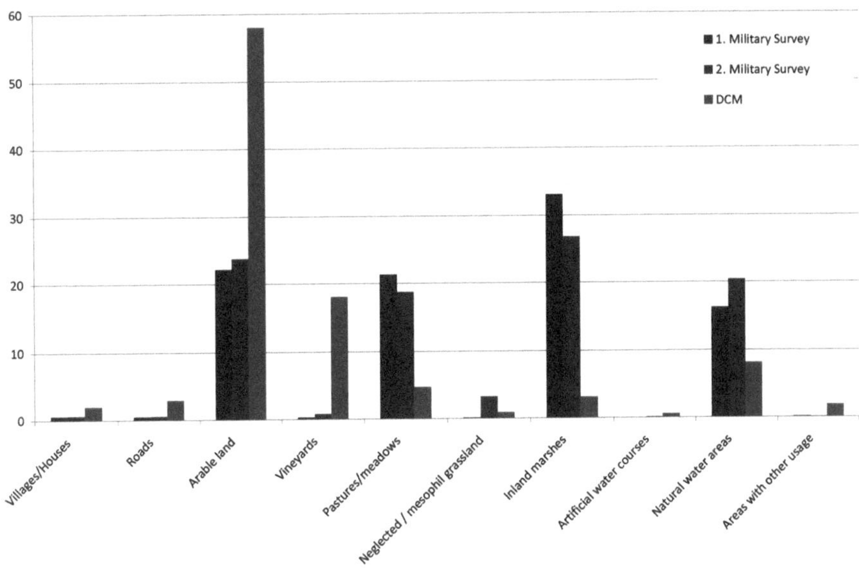

Fig. 9.2 Alteration of land-cover/use categories

9.4 Results

9.4.1 Interpretation

When looking into details of change of areas and their alteration a lot of characteristic landscape changes can be seen. The built up area was similar in both historical sheets but has increased three times (since the 1840s) and four times (since the 1780s) respectively up to the Twentieth century for instance (see Fig. 9.2). The traffic area enlarged five times (since the 1840s) and six times (since the 1780s) respectively of the original area, considering, that historical traffic areas were buffered 6 m wide but today's traffic areas are often much wider.

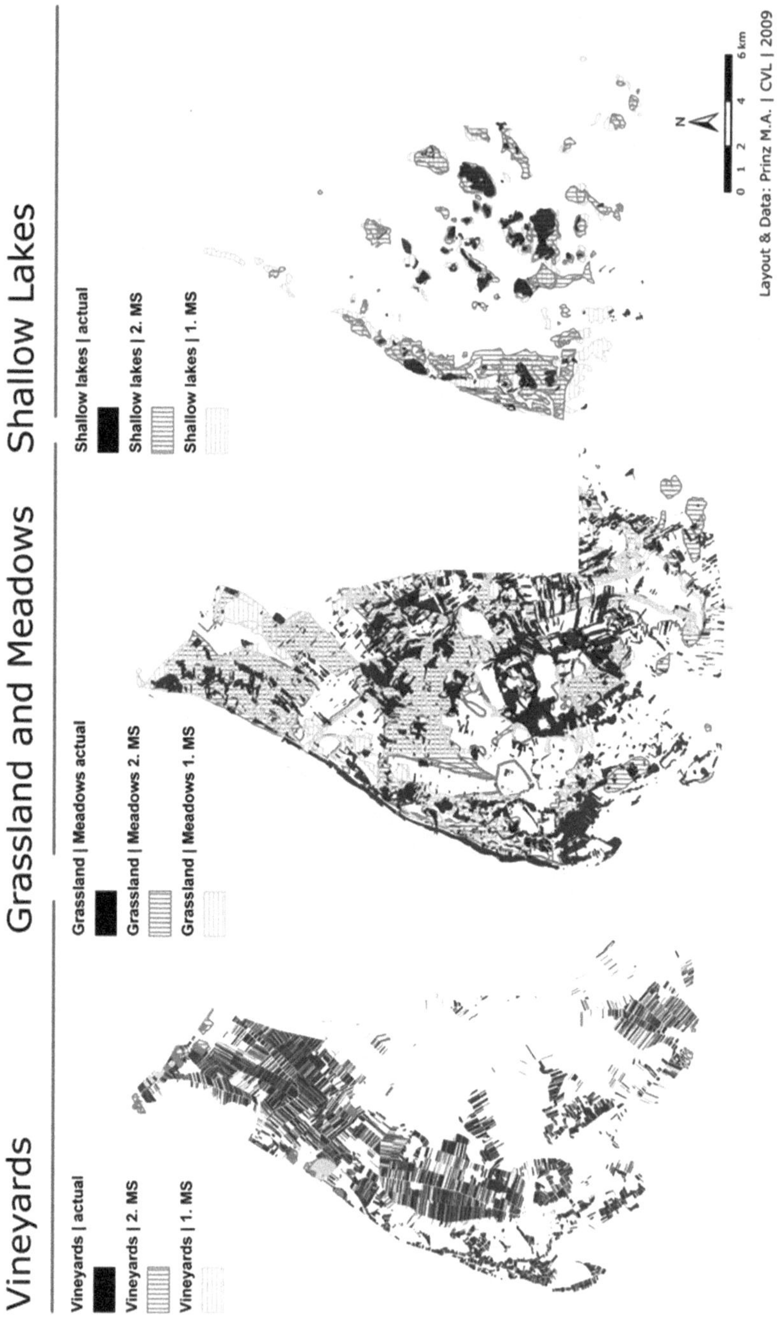

Fig. 9.3 Change of major land-use/cover types

Like in other favoured agricultural landscapes cropland has increased because of intensification of land use by transforming extensive rangeland into more productive land. Therefore, the respective area has been more than duplicated.

The most important change concerned viticulture. This former mostly small-scaled cultivation has increased to the second most common land-use system. Today's viticulture area is 78-times larger than in the eighteenth century and 23-times larger than in the nineteenth century. A contrary trend can be seen when looking at grassland. Here land-use area has decreased to a quarter (see Fig. 9.3). Even more dramatic is the loss of wetlands. Only 10% of the former inland marshes (areas dominated by sedges or reed) have remained (see Fig. 9.2). The area of shallow lake fluctuates but is now only half what it was in the eighteenth century (see Fig. 9.3).

9.5 Discussion

9.5.1 Thematic and Spatial Accuracy

When processing data from historical sources various problems occur. Sheet edges, which are useful contact points for georeferencing, were often damaged. Exact setting of reference points was often difficult and unavoidably sketchy. But a geographically correct positional arrangement of the interpreted data is essential for data analysis and illustration. This was only possible after a further treatment of data by time-consuming rubber-sheeting. Accuracy was then between 5 and 20 m which is much better as described by Cajthaml et al. (2007) for this kind of data.

One major problem was that individual sheets were created from different surveyors. This led to some inaccuracies in content: (1) Quality of used signatures differs from sheet to sheet. Only bad copies of some sheets were available. These sheets can be indicated as uniform in colour and with a bad resolution. (2) General accuracy was different between surveyors. Hence, in the western sheets of the 1840s wells are completely missing while in other sheets they are drawn meticulously. (3) Transition between sheets often causes abrupt land-cover change. This is probably because of different times of mapping (survey lasts decades – although only a single year has been quoted for the sheets) or probably because of different methods of operation. In general, sheets from the 1780s were of very good quality. Sheets from the 1840s from this area are however a conglomeration of sheets with different qualities and contents. An easy interpretation was therefore difficult over long passages.

9.5.2 Observed Landscape Changes

As in almost every agriculturally used landscape of the world, landscape is changing more or less dramatically, depending on the special landscape situation. Another analysis in Lower Austria shows an example where changes, especially concerning landscape structure, are less serious and historical and actual land cover and land use are rather equal (Lettner in prep.). In the Seewinkel, reasons for this broad change are intensification of agriculture and regulation of water level with drains and ditches (Draganits et al., 2006). This regulation can also be seen in increasing area of artificial water courses. Also reduction of open water courses appears along with this regulation and drawdown of groundwater level in the Seewinkel. This is especially relevant for ecological issues, for example occurrence and stability of shallow lakes associated with habitat quality of these lakes depend on the groundwater level. These dramatic changes led to one of the major problems of nature conservation today: appropriate areas and remnant biodiversity hot spots becoming smaller, more isolated and increasingly abandoned. These hot spots are always linked to special structures and types of landscapes and ecosystems respectively. Although more small conservation areas have a greater diversity then one large area (Higgs & Usher, 1980), undersized areas are more and more problematic. Especially undisturbed resting areas for migratory birds play a major role in the ecosystem of the Seewinkel (Steiner & Parz-Gollner, 2003).

9.6 Conclusions

To conclude, the former quadrinominal field-grassland-fen-water landscape has changed to a crop-viticulture landscape but still remnants of the former extensive land-use system are present. In case of a probable extensification of agricultural areas on marginal sites in a changing regional economic system these remnants could be a starting point for renaturation of parts of the area. The analysis of the DTM shows, that there will be enough additional potential lake areas for temporary filling-up. This potential will make it possible to re-establish a large and highly diverse connected landscape between Austria and Hungary in the future.

Acknowledgements We would like to thank Flora Hejas for translating the Hungarian signatures, Christa Renetzeder and Anna Herman for helpful discussions and the European Union and Land Burgenland for financing this study, which was part of the INTERREG-project "SISTEMaParc".

References

Abensperg-Traun, M., Wrbka, T., Bieringer, G., & Hobbs, R. (2004). Ecological restoration in the slipstream of agricultural policy in the old and new world. *Agriculture, Ecosystems and Environment, 103*(3), 601–611.

Antrop, M. (2004). Landscape change and the urbanization process in Europe. *Landscape and Urban Planning, 67*(1–4), 9–26.

Antrop, M. (2005). Why landscapes of the past are important for the future. *Landscape and Urban Planning, 70*(1–2), 21–34.

Baudry, J. (1989). Interactions between agricultural and ecological systems at the landscape level. *Agriculture, Ecosystems and Environment, 27*(1–4), 119–130.

Berger, R., Fally, J., & Lunzer, H. (1992). *Frischer wind am steppensee. Friedenspark im herzen Europas*. Deutschkreuz: Fally.

Bitenc, M. (2007). Analysis of airborne laser scanning data and products in the Neusiedler see project. *Ekscentar, 10*, 60–64.

Bossard, M., Feranec, J., & Otahel, J. (2000). *CORINE land cover technical guide – Addendum 2000*, Technical Report, EEA 40, 1–105.

Bürgi, M., Hersperger, A. M., & Schneeberger, N. (2004). *Driving forces of landscape change – Current and new directions. Landscape Ecology, 19*(8), 857–868.

Cajthaml, J. (2007a). Historické mapy středních měřítek – digitalizace a využití dat. In *Juniorstav. Odborná Konference Doktorského Studia* (pp. 355–357). Brno: VUT.

Cajthaml, J. (2007b). Georeferencing of Historical Military Mappings and Later Map Internet Publishing. *CTU Reports, Proceedings of Workshop 2007*, Czech Technical University, Praha, pp. 2.

Cajthaml, J., et al. (2007). Georeferencing and cartographic analysis of historical military mappings of Bohemia, Moravia and Silesia. *CTU Reports, Proceedings of Workshop 2007*, Czech Technical University, Praha, p. 2.

Crews-Meyer, K. A. (2004). Agricultural landscape change and stability in northeast Thailand: Historical patch-level analysis. *Agriculture, Ecosystems and Environment n., 101*(2–3), 155–169.

Draganits, E., et al. (2006). Neusiedlersee/Fertö Tó area (Austria/Hungary): landscape change through time constrained by geomorphology, lithostratigraphic sections, archaeology and historical maps. *Geophysical Research Abstracts, 8*, 1–3.

Dömötörfy, Z., Reeder, D., & Pomogyi, P. (2003). Changes in the macro-vegetation of the Kis-Balaton Wetlands over the last two centuries: A GIS perspective. *Hydrobiologia, 506*(1–3), 671–679.

Haber, W. (2004). Landscape ecology as a bridge from ecosystems to human ecology. *Ecological Research, 19*(1), 99–106.

Haeusler, H. (2006). The Parndorf- and Seewinkel gravel beds – Implications of Middle Pleistocene uplift and neighboring subsidence in the Northern Burgenland (Austria). *Geophysical Research Abstracts, 8*, 1–2.

Higgs, A. J., & Usher, M. B. (1980). Should nature reserves be large or small? *Nature, 285*, 568–569.

Jenerette, D. G., & Wu, J. (2001). Analysis and simulation of land-use change in the central Arizona – Phoenix region, USA. *Landscape Ecology, 16*(7), 611–626.

Kohler, B., Rauer, G., & Wendelin, B. (1994). Landschaftswandel. In G. Dick, M. Dvorak, A. Grüll, B. Kohler, & G. Rauer (Eds.), *Vogelparadies mit Zukunft? Ramsar-Gebiet Neusiedler See – Seewinkel* (pp. 21–36). Styria, Wien: Umweltbundesamt.

Kovács, G., Zámolyi, A., Székely, B., & Papp, S. (2008). Tectonically induced drainage system in the Western Pannonian Basin. *Geophysical Research Abstracts, 10*, 1–2.

Kozak, J. (2003). Forest cover change in the Western Carpathians in the past 180 years. A case study in the Morawa region in Poland. *Mountain Research and Development, 23*(4), 369–375.

Krejčí, J., & Cajthaml, J. (2007). Mullerovy mapy českých zemí, jejich digitalizace a zpracování pro internetovou vizualizaci. *Kartografické listy, 15*(2007), 51–59.

Kretschmer, I., & Riedl, A. (2008). Austrian (Habsburg) official cartographic activities from the First Military Survey to World War I. *Geophysical Research Abstracts, 10*, 1–2.

Křováková, K., & Brůna, V. (2007). Historical landscape of Šumava in the light of paleobotanic and antique maps' evidence. In R. G. H. Bunce et al. (Eds.), *25 years of landscape ecology: Scientific principles in practice*, Proceedings of the 7th IALE World Congress 8–13 July. Wageningen: IALE Publication Series.

Lettner, C. (in preparation) *Langfristiger (1822 bis 2006) Wandel einer Kulturlandschaft im Übergangsbereich nördliche Kalkalpen – Flysch*, Diplomarbeit, Universität Wien, Wien.

Löffler, H. (1982). *Der seewinkel – Die fast verlorene landschaft*. St. Pölten: Verlag Niederösterreichisches Pressehaus.

Mikšovský, M., & Zimová, R. (2007). Staré mapy Čech – vybrané aspekty kartografického jazyka. *Kartografické listy, 15*, 87–95.

Nationalpark Neusiedlersee-Seewinkel (2008) *Offizielle Homepage des Nationalparks Neusiedlersee-Seewinkel*, http://www.nationalpark-neusiedlersee-seewinkel.at/impressum.html

Naveh, Z. (1995). Interactions of landscapes and cultures. *Landscape and Urban Planning, 32*(1), 43–54.

Nelson, S. A. C., Soranno, P. A., & Qi, J. (2002). Land-cover change in upper Barataria Basin Estuary, Louisiana, 1972–1992: increases in wetland area. *Environmental Management, 29*(5), 716–727.

Nestroy, O. (2005) Introduction to Austria – a general geographic survey. In A. Aichberg & A. Dadora (Eds.), *Soil organic matter and element interactions* (pp. 1–1). Austrian-Polish Workshop. ALVA-Mitteilungen Heft 3.

Peterseil, J., Wrbka, T., Plutzar, C., Schmitzberger, I., Kiss, A., Szerencsits, E., et al. (2004). Evaluating the ecological sustainability of Austrian agricultural landscapes—the SINUS approach. *Land Use Policy, 21*(3), 307–320.

Steiner, W., & Parz-Gollner, R. (2003). Actual numbers and effects of recreational disturbance on the distribution and behaviour of Greylag Geese (Anser anser) in the Neusiedler See – Seewinkel National Park Area. *Journal of Nature Conservation, 11*(4), 324–330.

Tasser, E., Teutsch, A., Noggler, W., & Tappeiner, U. (2007). Land-use changes and natural reforestation in the Eastern Central Alps. *Agriculture, Ecosystems and Environment, 118*(1–4), 115–129.

Timár, G., & Pišút, P. (2008) Changes of the sinuosity of the Morava River (western Slovakia) and their neotectonic origin. *Geophysical Research Abstracts 10, A Series*, 3–3.

Timár, G., et al. (2007). Projection and datum parameters of the second military survey of the Habsburg Empire (1806–1869) for GIS data integration purposes. *Geophysical Research Abstracts, 9*, 1–3.

Timár, G., et al. (2008). Combination of historical maps and satellite images of the Banat region—Re-appearance of an old wetland area. *Global and Planetary Change, 61*(1–2), 29–38.

Umweltbundesamt (2004) *Ramsar in Österreich, Innsbruck*, http://ww.umweltbundesamt.at/

White, M. A., & Mladenoff, D. J. (1994). Old-growth forest landscape transitions from pre-European settlement to present. *Landscape Ecology, 9*(3), 191–205.

Wrbka, T., et al. (2002). *Kulturlandschaftsgliederung Österreich. Endbericht des gleichnamigen Forschungsprojektes*, Forschungsprogramm Kulturlandschaft 13 (CD-ROM), Bundesministerium für Bildung, Wissenschaft und Kultur, Wien.

ZAMG – Zentralanstalt für Meteorologie und Geodynamik (2002) *Klimadaten von Österreich 1971–2000*, http://www.zamg.ac.at

Chapter 10
Conditions of Living – Reality, Reflections, Comparisons and Prospects

Milan Jeřábek

10.1 Subjective Perception and Real Landscape

Geography, being a science that reflects on and respects both a holistic approach and a hierarchical system of classification, takes up a specific position in studies of local and regional developments. Besides standard approaches, so-called *soft methods*, usually used in sociology, have been gaining more importance. Not only do they enable scientists to verify data obtained through more traditional methods, but such sift methods also complement our objective knowledge with subjective data and self-reflection.

The development of a region or a locality is based, among other things, upon what active approaches the permanent population of the area has adopted. Kollár (2000) says that approaches, beliefs, values and the importance attached to the environment affect the emotional state of an individual as well as their ideas, which both modify their complex behaviour. It is exactly the perception of what is currently going on, no matter whether gained through personal experience or by media that constitutes a significant factor in the development of a community and society in general.

When designing suitable empirical enquiries, geographers can focus their questionnaires on the following groups of questions (Jurczek & Günter, 1994, p. 1): "What are the effects of changes in general conditions? Which areas (in regional development, of society or community) are affected the most and the least? What development opportunities or threats have to be taken into account in a particular locality or region? How do various target groups perceive the ongoing changes?"

The Ústí Region, or more specifically its coal-mining core area, is well known to the public as a territory which is industrially used in an intensive way, suffering from pollution and to a large extent also from considerable social problems. This state of affairs has stimulated, among other things, the profound scientific interest of various researchers and institutions. Naturally, the local university is no exception and, due

M. Jeřábek (✉)
Department of Geography, Jan Evangelista Purkyne University in Ústí nad Labem, České mládeže 8, 40096 Ústí nad Labem, Czech Republic
e-mail: milan.jerabek@ujep.cz

J. Anděl et al. (eds.), *Landscape Modelling*, Urban and Landscape Perspectives 8,
DOI 10.1007/978-90-481-3052-8_10, © Springer Science+Business Media B.V. 2010

to unhampered cooperation between its departments, the university has been able to adopt a *complex (interdisciplinary) approach*. Several completed projects such as Jeřábek (2006) dealing with spatial aspects, Zich (2006) with an emphasis on social matters or environmentally oriented, are documenting the wide scope of the approach.

It is not the intention either to present in this paper profound theoretical analysis or to describe the methodology of our empirical enquiries. The research rather concentrates on some selected data that mainly concern the general perception and evaluation of complex living conditions.

10.2 Methodology and Focus

There is a method which enables us to analyse reality quite well and which has been gaining more and more in importance: the *SWOT analysis*. This method describes current situations through indications of strengths and weaknesses both at a general level and in a broader context with specific focuses on either particular regions or particular aspects. The situation "inside" (i.e. endogenous potential) is complemented by external factors (agents) which are or will be affecting the development of the territory in question either positively (opportunities) or negatively (threats). In this particular case, the author has used the results of several students' diploma theses which dealt with selected territories and which can be considered as case studies.

In each *empirical enquiry*, it is absolutely vital to follow an order of logical steps (for more detail see Jeřábek & Anděl, 2005), namely the selection of a focus, the formulation of suitable questions, the definition of a target group, the fieldwork and finally the evaluation and interpretation of results.

The Geoscape Project in itself, supported by the Grant Agency of the Czech Republic and the Ministry of Labour and Social Affairs, is quite broadly based upon results concerned with a relatively wide range of aspects for *sociological surveys* of population. For comparison, there were asked 12 questions at the national level and 21 questions at the regional level. In this paper, only selected data indicating living conditions and socio-economic development will be presented.

At the national level, a professional agency was collecting the questionnaires (September, 2005; 1.045 respondents), and the regional survey was carried out in NUTS 3 Ústí Region, NUTS 3 Karlovy Vary Region (forming a NUTS 2 region North-West) and NUTS 3 Liberec Region (hereafter KV-UL-LB) with a total of 153 respondents. In addition, students of our university carried out enquiries in the Ústí Region in their optional courses and diploma theses (from 2005 till 2007, total number of 918 respondents in eight different areas). The structure of the respondent groups follows the common structure of the population, so the results can be considered as representative.

Besides the surveys among the population, *expert interviews with local mayors* (in 2006) were also carried out, the results of which are also used in this paper, but selected aspects have been considering successes and failures of the settlement in question since 1989 and current and future priorities and problems. *Intentions of regional authorities* were evaluated on the basis of their publication called Strategie udržitelného rozvoje Ústeckého kraje 2006–2020 [Sustainable Growth Strategy for the Ústí Region] (Hřebík et al., 2006), which listed measures and investments intended to be made in the next 15 years.

10.3 Perception of Living Conditions and Comparisons

10.3.1 Current Conditions of Living

In each of the observed territorial units (with a few exceptions) the respondents expressed a satisfaction with their living conditions: the figures are 2.17 in the CR and 2.25 in KV-UL-LB (on a four-grade scale). It is interesting that none of the surveyed areas in the Ústí Region showed the same trend. Generally speaking, respondents in peripheral territories (both in the borderland and the inland) were more satisfied than those living in the industrial, ecologically strained microregion, where the evaluation was as bad as 2.78.

Despite the obvious differences between particular surveyed areas, it was possible to indicate several common characteristics in the SWOT analyses (strengths or weaknesses).

The geographical situation appeared to be a strength even though most of the surveyed areas were located outside main development zones (poles and axes, see the figure/map). The respondents really appreciated both close distances to rivers (the Labe, the Ohře) and their area's position near the border which results in opportunities to cooperate with Germany.

The landscape and environment seemed to be perceived as the greatest assets in the surveyed areas in the Ústí Region (cf. Raška & Oršulák, 2009). Apart from the general appreciation of the landscape's high value as well as its natural and aesthetic potential, there were some specifically mentioned phenomena: the CHKO České středohoří (Protected Landscape Area of Central Bohemian Highland) or the planned CHKO Střední Poohří (Middle Ohře) as well as several attractive localities – such as the volcanic hill of Bořeň. On the other hand, the respondents did not fail to mention the air pollution caused by industrial plants and coal-burning power stations or unresolved environmental issues, such as wastewater treatment, solid-waste treatment and the needed shift towards gas in household heating.

The population and employment were perceived mainly as weaknesses as there were many negative factors: a continuing decrease in the number of inhabitants, a low level of education and low qualifications, an insufficient number of job opportunities, an above-average unemployment rate and a high rate of commuting to places outside the region.

The economy was mainly characterised in terms of dramatic differences between traditional industrial or mining areas and other areas that are mainly orientated on agriculture or possibly crafts. Both "orientations" are currently experiencing a crisis, which is basically caused by low competitiveness, on the one hand, and by unfavourable natural conditions, on the other.

The number and quality of community *service facilities* depends on the positions of surveyed areas' in the settlement structure. The most significant complaints of respondents pointed to insufficient shopping and servicing facilities, and limited assortment of goods in shops associated with higher prices. The density of settlements also impacts the effectiveness of the social and physical infrastructure (i.e. education system, wastewater treatment, etc.).

Most of the surveyed areas also suffer from a lower level of accessibility by *public transport* and from lower standards of roads and other means of communication. As they are situated on the SW-NE development axis of the coal-mining basin, both major towns are characterised by heavy traffic including heavy truck transits.

The regions' attractive landscapes and rich histories provide a great potential for *tourism*. Unfortunately, the low quality of catering and accommodation facilities, insufficient services, ineffective promotion making information about the regions hard to obtain, as well as the neglect of historic buildings and monuments, prevent full realisation of possible tourism activities.

Table 10.1 gives a detailed overview of the respondents' evaluation of living conditions. It offers 17 different aspects of life in various spheres. The respondents living in the mountains near the border, in the areas with a traditional industrial manufacturing base and in the areas with a developing economic base expressed the most satisfaction. On the other hand, the most criticism has been heard in the inland periphery areas that are mainly orientated towards agriculture.

Among the various aspects, the surroundings of the settlement and the landscape received the most approval from the respondents, followed by housing, conditions in restaurants, education/school system and environment. Subjectively, the worst situation concerned job opportunities and cultural life.

10.3.2 Perception of Future Development

In similar terms, the SWOT analysis has been applied to future prospects – in this case the part dealing with opportunities and threats.

The borderland situation of the area is perceived in a positive way, mainly due to the cross-border contacts it can offer. It is never considered as a negative condition.

The value of landscape can be improved through changes in its management with the ultimate goal to preserve its character, as well as through providing support to ecological family farming focused on traditional products. In contrast, there are quite a few worries about damaging the landscape through intensive mining (in the Central Bohemian Highland) and about harm to natural beauty spots from tourism (cf. Balej, Anděl, Oršulak, & Raška, 2008; Balej & Anděl, 2008).

Objectively, positive changes resulting in decreasing unemployment can be achieved through improvements in the education/qualification structure of the economically active population. However, it is more likely that emigration of qualified labour will continue and that unemployment rates will continue to rise. In some of

Table 10.1 Subjective perception of various aspects of living conditions (author's own survey, 2006–2007)

Area aspect	Libčeves	Petrovice	Třebenice	Benešov	Klášterec	Vejprty	Verneřice	KV-UL-LB	CZECH
Human relationships	2.12	2.14	2.06	1.60	2.66	2.24	2.36	1.21	2.21
Housing	1.71	1.40	1.45	1.70	1.71	1.81	2.27	1.16	1.78
Job opportunities	2.50	2.25	2.58	2.25	2.15	2.42	2.51	1.14	2.37
Shopping facilities	2.05	2.62	1.88	2.24	2.19	2.03	1.81	1.44	1.35
Social and health care	2.29	1.96	1.88	2.09	2.50	2.28	2.46	1.16	1.90
Education/school system	2.07	1.81	1.98	1.53	1.51	1.78	1.96	1.43	1.70
Children's free-time activities	2.23	2.32	2.34	1.68	1.74	1.97	2.40	1.14	1.92
Cultural life	2.54	2.12	2.32	1.89	2.19	2.10	2.75	1.13	1.86
Sports life	2.12	1.90	2.13	2.32	1.93	1.75	2.33	1.16	1.72
Conditions in restaurants	1.95	1.40	1.58	2.14	1.45	1.66	2.02	1.30	1.58
Other services	2.35	2.08	1.91	2.09	2.02	2.29	2.75	1.14	1.87
Roads	2.04	2.23	2.04	2.37	2.15	2.27	2.53	1.15	2.09
Public transport accessibility	2.78	1.60	2.49	1.55	1.89	2.22	2.68	1.13	2.02
Recreation facilities for visitors	2.29	1.63	2.16	1.95	1.78	2.02	2.33	1.26	1.78
Safety/law and order	2.21	1.92	2.46	1.81	2.38	2.37	2.41	1.17	2.32
Environment	1.80	1.50	1.84	2.03	1.68	2.06	1.81	1.30	1.84
Surroundings, landscape	1.59	1.44	1.65	1.46	1.21	1.70	1.84	1.28	1.70
Total average	2.15	1.90	2.04	1.92	1.95	2.06	2.31	1.22	1.88

the surveyed areas, the issue of unemployment is closely related to issues of the Romany minority.

Considering the economic sphere, the support of entrepreneurial activities, including incentives for foreign investors, is seen as the best option. The other suggestion is to utilise local sources of energy. However, there are more frequent pessimistic expectations, such as a decrease in purchasing power leading to lower standards of living, a decline in agricultural production and a possible collapse of gardening and orchards.

Shortage of finance for recovery and development of the communal infrastructure is perceived as a threat in three of eight surveyed areas in the Ústí Region, including economically developed ones.

In transportation, a hope of some improvement can be seen in plans for better communications which are just being discussed or possibly in the construction of a bypass. On the other hand, a further reduction of bus or train services could have a negative effect on accessibility of various places outside the region that offer better job opportunities.

Tourism tends to be considered as the "panacea" for regional growth. It is believed that the income from tourism will grow due to increasing numbers of foreign tourists, wider spread of agrotourism, support for spas, for traditional cultural, sports and social events or due to extended utilisation of the mountains (e.g. the ski resort at Klínovec).

The subjective perception of *development prospects* follows tendencies of past and current advancements (Fig. 10.1). While the figures at higher level units are almost identical (average value in the entire Czech Republic being 1.87 and in

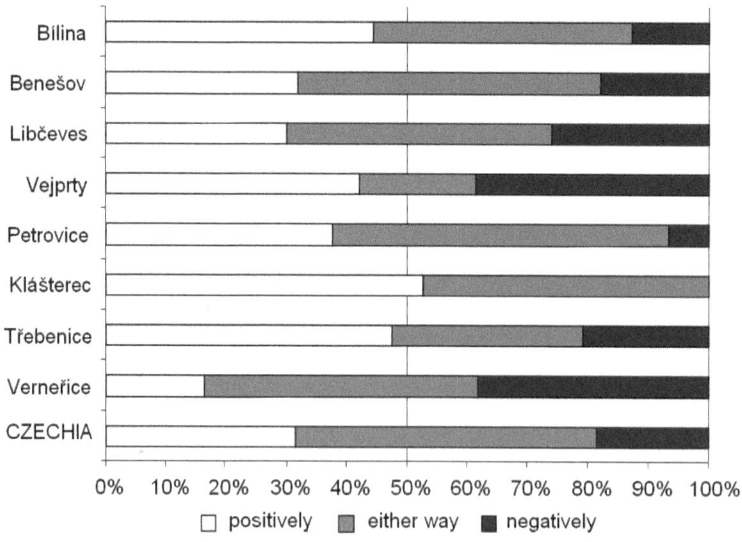

Fig. 10.1 Perception of the economic development in the model areas in the next 5 years (author's own survey, 2006–2007)

KV-UL-LB 1.84 on a three-grade scale), the surveyed areas display dramatically more variability. The regions that have been successful in a later development stage, i.e. a progressively oriented area, a borderland mountainous area and an agricultural area seem to have a successor: an industrial area that is environmentally strained. On the other hand, unfavourable developments are expected in the inland periphery as well as in the peripheral borderland zone microregion with difficult accessibility.

When evaluating the regional plans, the author considered the public's opinion of 20 specific priority intents in four different fields. Each of the priorities was marked on a school scale, i.e. 1–5.

It is significant that the respondents placed emphasis on the environment, as the increase in investments into environmental protection measures won in competition with 19 other plans. Moreover, improvement in air quality (reduction in air pollutant emissions) placed fourth. The second most approved priority is concerned with local economy, namely the support for current enterprises and environment-friendly technologies as well as support for small and middle-sized businesses oriented towards innovations (second respectively fifth place).

Among the other supported intentions there are also improvements in education/qualification structure (i.e. social sphere) and more environment friendly attitudes of local authorities (i.e. public administration). Surprisingly, the reduction in abortion and crime rates as socially undesirable phenomena appears among the least supported priorities. Other issues follow: the reduction in transportation/traffic

Table 10.2 Selected outcomes of interviews with local mayors (author's own survey, 2006)

Model area	Priorities and problems (current and future)	Rationales, explanations
Klášterec	Revitalisation program for prefab housing estates	A substantial proportion of housing facilities, unsatisfactory situation
Vejprty/Kovářská[a]	Settlement survival through preserving decent living conditions	Peripheral position, hard climatic conditions, insufficient services
Benešov and Verneřice[b]	Providing drinking water and wastewater treatment	Scattered settlements, insufficient infrastructure
Petrovice	Cycle path – interconnecting the Ore Mountain thoroughfare and the Elbe Sandstones	Focus on tourism including foreigners (Germans), attractive landscape
Bílina	A Chateaux – renting and reconstruction, extended cooperation	Centrally situated unused building in private hands, cultural events
Libčeves	Reconstruction of a chateaux vs. that of roads	Limited budget, history vs. current needs
Třebenice	Shortage of land for house construction	Pleasant environment, local authority's activities vs. unclear ownership

[a]The interview was not conducted with a representative of the core town.

[b]The interview was conducted at a meeting of a settlement association covering two model areas.

in the economic sphere, revitalisation programs for unused land and buildings as well as for the sanitation of old environmental damage, and reduction of debts of local authorities in the public administration sphere.

It is necessary to note that roughly one quarter of all respondents did not comment on these issues. The reason seems to be difficulty in commenting on something respondents have insufficient or no information about or on things that do not concern the respondent at all.

The selection applied to the interviews with surveyed area mayors is more qualitative than quantitative. Table 10.2 shows that *perceiving priorities and problems* is a highly individual matter which is very much dependent on specific conditions in the locality or region concerned.

10.4 Conclusions

The analysis presents only a part of the author's own empirical survey results. Despite this, it is possible to draw some conclusions concerning the survey areas. First of all, there is the wide variety of conditions on the territory of the Ústí Region in all aspects. The variety is represented by the survey areas having different endogenous (internal) potentials, their historic developments and geographical positions. In those areas, one can identify both common features and unique manifestations and problems. Naturally, exogenous (external) factors, such as general conditions set nation-wide, related legislation, effects of foreign capital inflow and, last but not least, regional policies of the European Union, play a significant role (Cappellin & Batey, 1993; Martinez, 1994; Bufon, 1998). This might be seen in several examples from Europe (e.g. Krätke, 1998; Matthiensen & Bürkner, 2001). It is possible to draw the following theses.

Satisfaction with quality of life in a locality of permanent address prevails. Basically, it has been confirmed that general satisfaction with conditions of living corresponds with specific satisfaction which was indicated through 17 aspects in Table 10.1. The respondents in the survey areas (eight areas in the Ústí Region) tend to be more critical about their living conditions than the respondents in higher level (comparative) units – both the combination of three regions (Karlovy Vary, Ústí, Liberec = KV-UL-LB) and the entire Czech Republic.

At the regional level, there is considerable satisfaction with the exception of shopping opportunities and the school system and education. The results at the national level are closer to those at the local level, and this is particularly true considering critical perceptions of job opportunities, safety, law and order, and road networks. Similarly, there is also close association of positive perceptions of surroundings and landscapes, conditions in restaurants, housing, and the school system and education.

The results of the empirical enquiry clearly show that there is a significant correlation between the perceptions of current situation and development prospects. Quite a few survey areas belong to more successful ones, but looking at the aspects considered, one can say none of them is characterised in clear terms. Among the more successful areas there are representatives of traditional craft areas, a representative with a significant centre situated on the edge of the Ústí Region, a representative of a borderline mountainous area drawing on cross-border activities and a representative of an agricultural area in the periphery. On the other hand, problems are only typically perceived in the mountainous area in the inland periphery.

Acknowledgements The research presented was supported by the project *Czech Borderland after Schengen: a Distinct, Oscillating and/or Transit Area?* (No. IAA311230901) funded by Grant Agency of the Czech Republic.

References

Balej, M., Andel, J., Orsulak, T., & Raska, P. (2008). Development of environmental stress in the northwestern part of Czechia: New approaches and methods. *Geografie–Sborník ČGS*, *113*(3), 320–336.

Balej, M., & Anděl, J. (2008). Land use changes and environmental stress accounting (case study from northwestern part of the Czech-German borderland). *Journal of Geography and Regional Planning*, *1*(5), 97–109.

Bufon, M. (1998). Border and border landscapes: A theoretical assessment. In M. Koter & K. Heffner (Eds.), *Region and regionalism No. 3: Borderlands or transborder regions – Geographical, social and political problems* (pp. 7–14). Lodz: University of Lodz and Silesian Institut in Opole.

Cappellin R., & Batey P. W. J. (Eds.). (1993). *Regional network, border regions and European integration*. London: Pion.

Hřebík, Š., et al. (2006). *The sustainable growth strategy for the Ústí region 2006–2020*. Ústí nad Labem in Czech): Ústav pro ekopolitiku.

Jeřábek M. (Ed.). (2006). *Regional research in the northwestern Czechia, Vol. 124, Studia Geographica VII. UJEP*, Acta Universitatis Purkynianae, Ústí nad Labem (in Czech).

Jeřábek, M., & Anděl, J. (2005). Sociogeographic and sociologic research of selected aspects of a cultural landscape in relation to the exploitation of informational datasets. In M. Balej & J. Anděl (Eds.), *Complex geographical research on a cultural landscape I* (pp. 75–86). Ústí nad Labem (in Czech): MINO.

Jurczek, P., & Günter, K. (1994). *Auswirkungen der Grenzöffnung auf Ober- und Mittelfranken – Eine Bestandaufnahme auf der Basis empirischer Erhebungen, Vol. 23, Kommunal- und Regionalstudien*, Carl Link Verlag, Kronach-München-Bonn.

Kollár, D. (2000). Slovenská migrácia za prácou do Rakúska – realita verzus predstavy. *Geografie – Sborník ČGS*, *105*(1), 41–49.

Krätke, S. (1998). Problems of cross-border regional integration: The case of the German-Polish border region. *European Urban and Regional Studies*, *5*(3), 249–262.

Martinez, O. J. (1994). The dynamics of border interaction. New approaches to border analysis. In G. Blake (Ed.), *World boundaries I* (pp. 1–15). London: Groom Helm.

Matthiensen, U., & Bürkner, H. J. (2001). Antagonistic structures in border areas: local milieux and local politics in the Polish-German twin-city Gubin/Guben. *GeoJournal*, *54*(1), 43–50.

Raška, P., & Oršulák, T. (2009). An overview of natural risks in the Ustecko region. In J. Anděl et al. (Eds.), *Complex geographical research on a landscape II* (pp. 93–112). Ústí nad Labem (in Czech): MINO.

Zich F. (Ed.). (2006). *Man in Borderland*. Ústí nad Labem (in Czech): UJEP.

Part V
Modelling and Geovisualisation in Landscape Planning and Management

How can geoinformation technologies help in participatory landscape and regional planning? How did the urban landscape of small towns change during the last 300 years? Why is there variance in the ways landscape and urban planning is appreciated in different groups of the decision-making triangle? What are the tools for identification of landscape structure changes? How do the geostatistical tools articulate the internal and interregional landscape diversity? What are the challenges of GIS in assessment of biodiversity in protected areas?

Chapter 11
Geovisualisation of an Urban Landscape in Participatory Regional Planning

Tomáš Oršulák and Pavel Raška

11.1 Introducing the Geoinformatics Approach to Urban Landscape

Efficient and optimal landscape development is a primary theoretical and practical interest of geographers and specialists from the decision-making sphere. However, such development requires understanding of general laws and concepts operating in the landscape and also existence of quality information as a basis for the decision-making process. From the point of view of general conceptions it is mainly important to emphasise that landscape is perhaps the most suitable criterion for studying the interactions of global and local structures and processes (Hobbs, 1996), as it reflects internal heterogeneity sufficiently and at the same time also the structural and development particularity compared to the other landscapes (cf. Hampl, 1971). However, it is also clear that every type of landscape requires (apart from an understanding of general principles) a different approach as regards the data and methodological basis as well as the different ways of impressing environmental sustainability on it.

A specific type of landscape is an urban landscape (sometimes also called urbanised landscape), the hard to understand complexity of which is also clear from the variety of approaches to its (still) geographic study (Oršulák, Raška, & Suchevič, 2007). If we see urban landscape research as a study of interactions of society and nature, the range of conceptions and methods is highly differentiated and in many cases it only emphasises one of the components, which may be caused by the fact that the concept of nature in towns was traditionally understood as an oxymoron (Hough in Meyer, 2005). For a long time, studies of towns maintained their social or political geographic nature (e.g. Sýkora, 1994; Hall, 2006). This is based partially on the urban ecology which was constituted in the 1920s in the so-called Chicago school. On the contrary, emphasis was placed on the natural elements in Central Europe where a field with a similar name was formed – town

T. Oršulák (✉)
Department of Geography, Jan Evangelista Purkyne University in Ústí nad Labem, České mládeže 8, 400 96 Ústí nad Labem, Czech Republic
e-mail: tomas.orsulak@ujep.cz

J. Anděl et al. (eds.), *Landscape Modelling*, Urban and Landscape Perspectives 8,
DOI 10.1007/978-90-481-3052-8_11,

ecology (*Stadtökologie*) focused on specific ecological features of town landscapes in relation to societal pressures and needs (Sukkop & Wittig, 1993). And thus while geographers often used the terms landscape and environment when they were trying to argue for a unity of geography by means of the integrated (but in reality often missing) scientific approach, it was rather the experts from related fields who contributed to the more complex understanding of the urban landscape (e.g. Gibbs & Healey, 1997; Haken & Portugali, 2003; Williams, 1994; Longley, 2006). When we say more complex, we mean heterogeneity of approaches applied to the unified type of geographic object/process but not to its concrete representation (i.e. a concrete place). The large extent of the studied cities often prevented integration of various methods of study and thus one or more approaches were preferred, which is summarised at the historic and scientific level by Hall (2006), i.e. from the study of the physical environment and morphology of towns, from the positivistic and behaviouristic approach to the structural and urban sociologic approach. In this particular case it is also quite unsuitable to use methods of multi-temporal land-use analyses, which are otherwise beneficial. This is because the changes of land-use are conditioned by many inseparable and often non-identifiable factors (driving forces), which can be interpreted only with difficulty within the scope of cities, resp. towns. For this reason, the majority of such works also focus their attention on complex, rural and possibly also suburban areas (Bičík, Jeleček, & Štěpánek, 2001; Bičík & Kupková, 2006). In this respect it is more suitable to study small towns for instance by works concerning regional development and environment (Anděl et al., 1992; Vaishar et al., 2001).

In the last decades, the complex study of towns has been assisted by information technology, which exponentially increase its performance reflected in the possibilities of visualisation of past, contemporary and possible future landscapes. The necessity of such continuity of past – recent – future landscape development in regional and landscape planning was well explained by Antrop (2005). These visualisations of landscape development have various forms and they are based on the use of various data sources. For example, for the purposes of historic landscape visualisation, military maps from the mid-eighteenth to mid-twentieth century, and photographs and aerial photographs from the mid-Nineteenth century are used. Various paintings, drawings and engravings depict the landscape even before the use of photographs began but their usability as regards authenticity is disputable. The condition of the contemporary landscape can already be reflected and depicted with a significantly higher degree of credibility, thanks to regular taking of aerial photographs and significantly more exact data concerning the landscape as such. Construction of the potential future landscape condition or rather scenarios of landscape development can be a suitable tool for community/participatory planning (Tress & Tress, 2003; Buchecker, Hunziker, & Kienast, 2003) including the modelling issues (Walz et al., 2007), but at the same time it runs up against various problems that can be partially overcome by means of objective and fixed points the most important of which – in the Czech planning community – undoubtedly is a territorial plan. On the other hand we can ask: who creates the territorial plan? Are these politicians or planners? Can we forecast any decisions concerning changes to the territorial plan? Can we guess the time in which the planned changes will take place? For these questions it is clear that visualisation of the future landscape is ambiguous and it has a dynamic development in time, it changes both on the basis of objective reasons and also subjective decisions. Is there any point in the modelling of future scenarios and their presentation in the form of visualisations? The answer is not easy and definite, but still the reactions and interest from the decision-making sphere in these matters confirm to us that the answer is yes. Similarly, inquiry among inhabitants shows that they would like to have knowledge about future development and the opportunity to influence that development (see Discussion and Conclusions at the end of this article).

Leading research establishments both in the world and in the Czech Republic are engaged in monitoring of landscape development and its visualisation. In the book *Virtual reality in geography*, Lovett et al. (2002) published a study dealing with visualisation of a sustainable farming landscape from which it is clear that at the technical level GIS and VRML (virtual reality markup language) are suitable tools for landscape visualisation. The development in the information technologies supports and accelerates the development of the whole area, such as 3D landscape visualisation, and specifically urban landscape visualisation (Batty, 1996; Spradley & Welch, 1998).

In the submitted work, the authors deal with the issues mainly from the angle of possible use of reconstruction geovisualisations for regional planning of urban settlements and landscape with emphasis placed on three aspects: (a) interrelation of past and future landscapes, (b) complexity of visual representation as a tool for optimal and sustainable planning, (c) involvement of inhabitants in the planning process using the assessment of subjective perception of photo-realistic and virtual-reality scenarios. In a case study the town of Klášterec nad Ohří is then focused on. The selected angle requires a complex approach, i.e. from the building of a database to the final landscape visualisation and from socio-cultural aspects to connections of the regional development with the natural environment. It is thus a complex approach typical for geography, in the particular case including structural, developmental and methodological complexity (cf. Hampl, 1971, 1998).

At the general level, the technical focus of this chapter is on monitoring of development of the urban landscape, its assessment, submission of measures, proposing of changes and actual visualisation of particular changes in the landscape. Specific partial targets were quantification of changes of the spatial growth of the model time in time horizons with regard to relatively stable factors. The quantification results were also interpreted as regards the long-term continuity of the local landscape development and with regard to the current strategic plan of town development.

11.2 Case Study: Mid-Size Town with Abrupt Shifts in Development

11.2.1 Study Area

The town of Klášterec nad Ohří is a part of a wider model area called Klášterecko. Klášterec nad Ohří is situated in the western part of the Ústí nad Labem region and the area of the town cadastre is approximately 5.500 ha. The town's population is almost 16.000. The town is divided into 12 parts from which Klášterec nad Ohří and Miřetice u Klášterce nad Ohří were selected for reconstruction.

The model settlement was selected on the basis of several criteria. In the monitored locality we have to be able to identify the studied factors (see below) in the time horizons (1750, 1850, 1964, 2000 and future). Another criterion was assumption of dynamic and heterogenic settlement development (i.e. spatiotemporal developmental non-linearity), or landscape transformation determined by various motive forces which in some cases enabled preservation of the historic features of the settlement while in others they were concealed (cf. Anděl, Jeřábek, & Oršulák, 2004; Balej & Anděl, 2008).

11.2.2 The Decision Triangle

We propose the term decision triangle as a representation of conditioned consensus among three groups involved in the process of landscape planning. These groups are representatives of the town (city), its inhabitants and finally specialists in landscape planning. Each within the framework of these groups has its own interests and preferences and, of course, is not homogenous in its opinions (differences are conditioned by education, religion, social position and many other factors). Town representatives see their decisions of complex economic and social development of the town, whereas the inhabitants prefer a viewpoint based on living and working conditions, and town planners prefer long-term and extensive projects, which are often considered apart from ecological (physical) features, processes and flows in a landscape. Planners are often pushed into unsuitable and unsustainable projects by politicians who – thanks to the relatively short term of office – need to present the prompt, "visible" and popular results of their decisions.

The decision triangle should, therefore, be complex, acting and deciding together on the basis of real landscape development scenarios. Moreover, such a mutual communication should include, besides the assessment of a concrete plan, the evaluation of direct and indirect project costs.

11.2.3 Database Creation

A database for subsequent geostatistical analysis, geovisualisation of the historic landscape and alternative scenarios of town development was created parallel to the selection of the model area. As regards gaining of topographical information, the most important data sources were archive aerial photographs and cadastral maps. The altitude data were obtained from a digital model of the territory (DMU 25) and subsequently corrected in some locations by GPS measurements. The thematic data, e.g. concerning the use of the land plots, buildings, quality and type of vegetation, were obtained from field surveying and use of secondary sources (historic maps, land registers, photographs etc.).

Mapping of land use and land cover requires interpretation of aerial photographs (in our case also of historic maps) and implementation of vectorisation of the grid content in the GIS environment.

For interpretation of land use we selected a method of visual photo-interpretation, which proved to be the most exact for detection of changes on the basis of aerial photographs. Despite its speed, both the controlled and uncontrolled classification was inaccurate mainly due to incorrect determination of areas in the shade and due to the need for post-classification adjustments. On the other hand, the process of visual photo-interpretation is demanding as regards personal experience and knowledge of the territory (cf. Paine & Kiser, 2003).

Vectorisation was implemented in software from ESRI, specifically ArcView, which is a part of ArcGIS. A data model was created that contained layers identical with the categories specified above and the relevant attributes corresponding to the particular category (e.g. for routes the attributes will be: type, length, ID). Concretely, the following thematic groups were interpreted: vegetation cover (eight categories), construction (four categories), water (two categories) and routes (two categories).

For some needs, the selection of elements was generalised into eight categories (forest, arable land, permanent grass areas, construction, water bodies, watercourses, roads, railways); the attributes remained the same. In comparison with for example CORINE 2000 (area divided into five basic categories and 15 subcategories) or with MUC – Modified UNESCO Classification (ten basic categories), our division of the area is simpler, mainly due to the impossibility of obtaining historic data concerning some types of areas.

11.2.4 Assessing the Spatial Matrix of Historical Growth

The regional development of the studied settlement was monitored in relation to the selected natural and socio-geographic criteria which were understood as relatively stable conditions, resp. factors valid for all time horizons. The first factor was slope inclination (S), which was divided into intervals based on the method of assessment of morpho-lithologic systems and their suitability for construction (Stankovianski, 1992). Only the boundary value of 2° was added due to the necessity to structure the first interval. The other factors included altitude (A) and distance from water courses (DWC), from the socio-geographic factors it was distance from railway (DRW) and distance from the historic centre (DHC), which is represented by two poles – a square and the castle.

Within the scope of geostatic analysis, these factors as well as the GIS theme of temporary construction changes are shown as GRID layers, in our case with a field size of 25 × 25 m. The representation of fields (pixels) in the assessed time horizons together with the information concerning the territorial extent of the town is shown in Table 11.1.

As regards factors requiring height (z) information, its reliability in the selected localities was checked by field survey (GPS) and comparison with analogue maps.

Table 11.1 Data sources and representation of assessed time horizons as regards area

Year	Pixels (n)	Pixels (f %)	Pixels (F %)	Basic data source
1750	28	1.65	1.65	Old map (I. military mapping)
1850	6	0.35	2.01	Old map (II. military mapping)
1964	374	22.10	24.11	Aerial photo
2000	417	24.65	48.76	Aerial photo
Future	867	51.24	100.00	Strategic plan of development

Years of first two sources are approximated to horizons for other statistical assessments.

The actual statistical analysis was performed by means of map algebra where for various intervals of the selected factor and also for time horizons of construction development we allocated values (weights) by means of the simple mathematic processing of which it is possible to express a unique resulting value for representation of each of the factors in the construction in the particular time horizon.

The results of geostatistical analysis of connection of the spatial development of the town with the selected natural and socio-geographic factors and conditions are represented in the graphs in Fig. 11.1 and they show changes of dependence of the location within the scope of the assessed time horizons.

The results divide the regional development of the model settlement into two longer stages. In the first stage (1750 and 1850) the construction is mostly localised in two or maximum three intervals of the relevant factor, while in the second stage represented by the time horizons of 1964, 2000 and Future, the range of intervals is wider. However, the width of interval range does not mean here a simple trend towards disadvantageous conditions (e.g. steep slopes). Its reason is rather less concentrated regional development in more directions with different conditions. An example is the condition (criterion) of distance from water courses (DWC), where in the first stage the intervals of 0–200 and 201–400 m prevail, and water courses are an important localisation factor for development (both residential and economic). On the contrary, in the second stage of development, in accordance with the decreasing importance of the natural localisation factor, which is quite a typical feature of transition to the post-industrial period, the development is situated in all intervals (max. 1.001–1.200 m). Even more distinct heterogeneity of the range of assessment criteria intervals can then be seen as regards distance from railway (DRW) and distance from the historic centre (DHC).

11.2.5 From Reconstructive Geovisualisation Towards Landscape Scenarios

Spatial data visualisation (SDV), sometimes also called geographic visualisation (or geovisualisation) is part of every study focused on geography. The method of urban landscape reconstruction was discussed in previous works by the authors

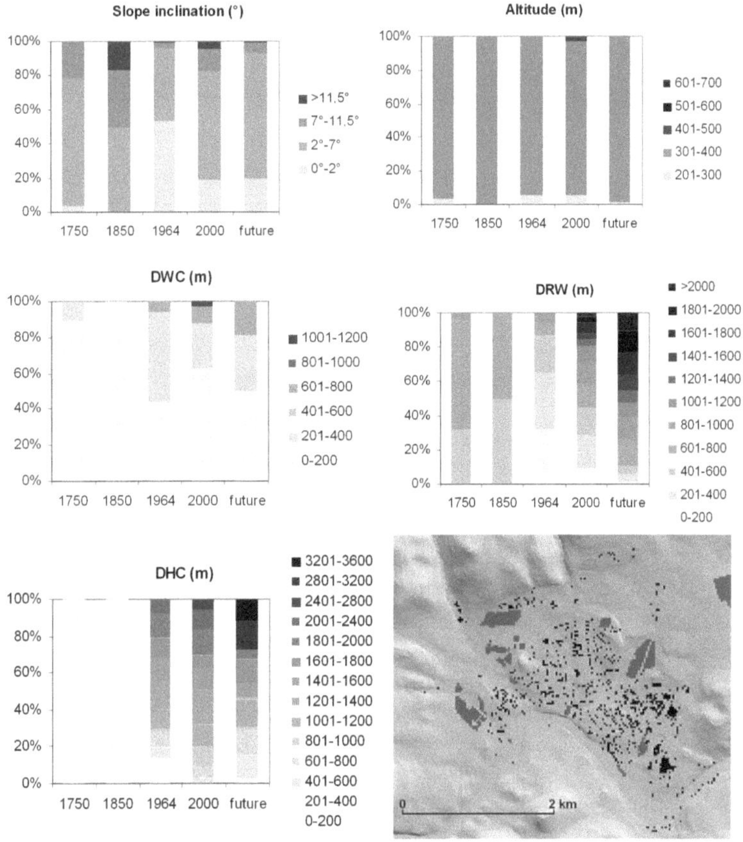

Fig. 11.1 Dependence of territorial growth of the town on various conditions/factors. In the *right-hand bottom* corner – initial GRID layer of territorial growth of Klášterec nad Ohří (see text for explanation of abbreviations)

(Oršulák et al., 2007; Raška, Oršulák, Anděl, & Balej, 2007), which are focused on creation of so-called hyperdata, i.e. multi-temporal data of various nature (analogue and digital map sources, image and statistic documentary sources etc.) and their processing, mutual correction and increased reliability and validity.

For the purposes of interpretation, data were visualised in a form of 2D maps in the ArcMap environment from ESRI. The resulting alternative scenarios of the future landscape and historic reconstructions were loaded on the TIN structure created from the altitude element of the digital model of the territory (DMU 25), which was updated and made more accurate by GPS measurement. To get an idea of the development of the settlement structure, a reverse reconstruction was used (Fig. 11.2). In the first stage, a 2D model of the existing development was prepared on the basis of ortho-photographs and field surveying. With the use of the historic data, the newer development was excluded gradually starting from the present

Fig. 11.2 Reconstructive geovisualisation of the Klášterec nad Ohří town. *Black* – Eighteenth century, *grey* – Nineteenth century, *white* – Twentieth century

and continuing to the past or any destroyed development was complemented. As regards any disputable parts of the development, field surveying proved the age of the building and its inclusion into a particular time horizon.

Visualisation of development within the scope of alternative scenarios of future landscape was accurate when for non-existent development we used a territorial plan and strategic plan of development of the town of Klášterec nad Ohří.

In the end, mainly for the purposes of more intuitive understanding of space, as is mentioned for example by Kraak et al. (2002), a 3D model of the town was created in the particular time horizons. As regards geometry, 3D models of buildings comprise of ground plan, information concerning height and roof shape. The ground plan of buildings was obtained from the above-mentioned vectorisation of ortho-photographs and for non-existent historic buildings from cadastral maps. The height and in some cases also the roof shape was gained through a method described for example by Suveg and Vosselman (2000), when the ground plan of buildings is divided into elementary geometric shapes and they are allocated one of three roof types (a flat roof, a gable roof and a hip roof). In a 3D model, the actual vegetation was replaced with the corresponding 3D image of the individual vegetation elements, for which a 3D Dosch database was used. The resulting models were exported into VRML format, for static 2D images we used JPEG format.

On the basis of the implemented reverse reconstruction and field surveying, we created a 3D model of the urban landscape of Klášterec nad Ohří, the visualisation of which can be seen for instance in Raška et al. (2007). The grey tones are used to depict the parts of development corresponding to the particular time horizon. For visualisation of an alternative scenario of the urban landscape we used two approaches. In the first case we used a 3D model and we represented the changes by means of 3D primitives, in the second we created photo-realistic views of the individual changes in the landscape (see e.g. Colwell, 1997) on the basis of the data specified in the territorial plan and from the field survey (Figs. 11.3 and 11.4). The

Fig. 11.3 Panoramic view of the area north-west of the swimming pool intended in the official land-use plan for construction of low-rise houses: (**a**) current situation; (**b**) photo-realistic alternative scenario with low-rise houses.[1]

reconstruction model of the urban landscape was exported to VRML format. The VRML format has shown itself to be the most suitable especially for its universality as well as the ability to present the 3D model in the World Wide Web environment. In this way it is simple for all the involved groups to access the landscape scenarios. This is especially true for inhabitants, who are often excluded from the planning process (see The Decision Triangle).

Fig. 11.4 View of the planned zone of urban-type development at the approach to Klášterec nad Ohří from Karlovy Vary: (**a**) current situation; (**b**) photo-realistic alternative scenario of low-rise terraced houses.[2]

11.3 Discussion and Conclusions

The reconstruction of a landscape representing the last 250 years represents a basis for geostatistical analysis of town development and its connections to selected conditions/factors and further it may be used for the assessment of future changes in the study area according to the territorial plan and strategic plan of development. However, when interpreting the results of geostatistical analysis, we have to consider the role which is played here by high dynamism and multi-factor interdependence of social systems development and urban systems particularly (cf. Hall, 2006; Sýkora, 1994; Bičík & Kupková, 2006). In the concrete case of settlements it means that in various periods different localisation factors connected with processes of more general nature (urbanisation, suburbanisation, reurbanisation) play their role, as well as the fact that social and economic dependence of regional growth is influenced by a number of complex, mutually interconnected driving forces (e.g. employment, leisure time activities, cultural preferences, architectonic design etc.).

Another reason, which compared to the previous argument calls for greater structure being given to regional growth of the model settlement is "decreasing space". If an area of a certain type (as regards assessment criteria) was developed in the past, it could not have been used in the following periods, unless the former objects had been removed. However, determination of such areas is very demanding and often unfeasible, as it is necessary to analyse a large amount of inconsistent graphical and statistical text documents which are not even available for some periods.

Since the method is proposed primarily for the decision-making triangle (see above), we have tested its convenience for the individual groups of this triangle using the (a) sociologic inquiry, (b) interview with town mayor and further (c) the structured interview. The first two were aimed at determination of major positives and problems in the town development, while the third method of inquiry (structured interview) focused on a potential application of the methods proposed by authors as a tool for landscape planning.

The results of the sociologic inquiry were taken in the scope of the research project and comprised 151 respondents and moderated discussion with the town's mayor (in the year 2006). The results indicating the factors of both positive (landscape, environment, housing) and negative (problem with procural of business buildings) perception of the town by its inhabitants, as well as former successes (reconstruction of the square), future visions, demands (revitalisation of prefabs) and problems (traffic bypass) mentioned by the mayor seem to represent the ideal occasion to apply the above-mentioned methods to enable qualified decisions regarding the developmental and structural complexity of the town's landscape.

The possible implementation of methods presented in this chapter to landscape planning was verified using the structured interview with representatives of each of the three groups of the decision triangle. The results of these interviews divided all interviewers into two groups. The first group included the inhabitants and part of the town representatives, while the second (smaller) one comprised the planners and the rest of the town representatives. The first group appreciated especially the reconstructive geovisualisation and 3D visualisation as an effective planning tool. The second group does not perceive the method as wholly useful since (according to this group) it is not suitable to base future development on past development (compare this opinion with arguments in Antrop, 2005).

Generally, the creation of multi-dimensional alternative scenarios of landscape development seems to be an effective tool for landscape planning at least for some groups of the decision triangle. It can also be said that the relation between lifelikeness of the model and its "communication" with the decision triangle is linear. The question to resolve is the technical matter, as the closer one comes to an expression of reality the higher are the technical demands. On the basis of our case study, the representatives of the Klášterec nad Ohří town plan to present the block 3D model on its web site and in the long term also the photo-realistic one.

The reconstruction of an urban landscape as a basis for regional development has its rationalisation over several levels. In the first one, it enables us to conceive of development of the town as a historically organic system. In this respect, the future development is a continual representation of changes, which do not act discordantly, though they may partially overlay the past design of the town. Through this approach, the future changes only remove or transform the unfunctional objects or space and try to preserve or renew the natural and socio-cultural phenomena which are relatively stable over the long term. In the case study, these phenomena are represented for instance by the locations assuring the ecological functions of the town (e.g. stable fragments of greenery; see Sukkop & Wittig, 1993) or by the historic buildings and urbanistic complexes (Oršulák et al., 2007).

Acknowledgements The study is one of the partial outputs of the research project of the Ministry of Labour and Social Affairs of the Czech Republic (No. 1 J 008/04-DP1).

Notes

1. Such a photo-realistic scenario may help inhabitants of the territory to imagine the intended future design of a landscape and to express their opinions about "inadequate" changes of their surroundings.
2. The low-rise terraced houses will negatively affect the traditional design of the local landscape. As in the case of Fig. 11.3, the scenario may help to avoid unsuitable and unwanted changes in a territory.

References

Anděl, J., et al. (1992). *Generel životního prostředí Klášterce nad Ohří*. Ústí nad Labem: Výzkumný ústav výstavby a architektury.

Anděl, J., Jeřábek, M., & Oršulák, T. (2004). *Vývoj Sídelní struktury a obyvatelstva pohraničních okresů Ústeckého kraje*, vol. 88, Studia Geographica IV, Acta Universitatis Purkynianae, UJEP, Ústí nad Labem.

Antrop, M. (2005). Why landscapes of the past are important for the future? *Landscape and Urban Planning*, *70*(1–2), 21–34.

Balej, M., & Anděl, J. (2008). Land use changes and environmental stress accounting (case study from northwestern part of the Czech-German borderland). *Journal of Geography and Regional Planning*, *1*(5), 97–109.

Batty, M. (1996). Visualizing urban dynamics. In P. Longley & M. Batty (Eds.), *Spatial analysis: Modelling in a GIS environment* (pp. 297–320). Cambridge: GeoInformation International.

Bičík, I., Jeleček, L., & Štěpánek, V. (2001). Land-use changes and their social driving forces in Czechia in the 19th and 20th centuries. *Land Use Policy*, *18*(1), 65–73.

Bičík, I., & Kupková, L. (2006). Changes of land use in Prague urban region. *Geografie-Sborník ČGS*, *111*(1), 92–114.

Buchecker, M., Hunziker, M., & Kienast, F. (2003). Participatory landscape development: Overcoming social barriers to public involvement. *Landscape and Urban Planning*, *64*(1–2), 29–46.

Colwell, R. N. (1997). History and place of photographic interpretation. In W. Philipson (Ed.), *Manual of photographic interpretation* (pp. 3–48). Bethesda: American Society of Photogrammetry and Remote Sensing.

Gibbs, D., & Healey, M. (1997). Industrial geography and the environment. *Applied Geography*, *17*(1), 193–201.

Haken, H., & Portugali, J. (2003). The face of the city is its information. *Journal of Environmental Psychology*, *23*(4), 385–408.

Hall, T. (2006). *Urban geography*. London: Routledge.

Hampl, M. (1971). *Teorie Komplexity a Diferenciace světa*. Prague: Charles University.

Hampl, M. (1998). *Realita, Společnost a Geografická Organizace: Hledání Integrálního Řádu*. Prague: Charles University.

Hobbs, R. (1997). Future landscapes and the future of landscape ecology. *Landscape and Urban Planning*, *37*(1), 1–9.

Kraak, M. J., et al. (2002). Some aspects of geovisualization. *GeoInformatics*, *5*(1), 26–37.

Longley, P. (2006). Grand challenges, environment and urban systems. *Computers, Environment and Urban Systems*, *30*(1), 1–9.

Lovett, A. A., Kennaway, R., Sunneberg, G., Cobb, D., Dolman, P., O'Riordan, T., et al. (2002). Visualizing sustainable agricultural landscapes. In P. Fisher & D. Unwin (Eds.), *Virtual reality in geography* (pp. 102–130). London: Taylor and Francis.

Meyer, W. B. (2005). The poor on the Hilltops? The vertical fringe of a late nineteenth-century American city. *Annals of the Association of American Geographers*, *95*(4), 773–788.

Oršulák, T., Raška, P., & Suchevič, S. (2007). Rekonstrukční vícerozměrná geovizualizace městských krajin: příkladová studie a perspektivy. *Historická geografie (Historical geography)*, *34*(1), 334–350.

Paine, D. P., & Kiser, J. D. (2003). *Aerial photography and image interpretation*. New York: John Wiley & Sons.

Raška, P., Oršulák, T., Anděl, J., & Balej, M. (2007). Creating a visual historical perspective for sustainable development of urban landscape. In E. Ritschelová & E. Sidorov (Eds.), *3rd international conference on environmental accounting and sustainable development indicators – book of proceedings* (pp. 123–127). Prague: UJEP – UK.

Spradley, L. H., & Welch, R. A. (1998). The challenges of a 3-D modelling in a dense urban environment. In D. Fritsch, M. Englich, & M. Sester (Eds.), *ISPRS commission IV symposium on GIS – between visions and applications* (pp. 594–596). Stuttgart: ISPRS 32.

Stankovianski, M. (1992). Hodnotenie stavu prírodných a prírodných-antropogenných morfolitosystemov (na príklade vybranej časti Bratislavy). *Geografický časopis*, *44*, 174–187.

Sukkop, H., & Wittig, R. (Eds.). (1993). *Stadtökologie*. Stuttgart – Jena – New York: Gustav Fischer Verlag.

Suveg, I., & Vosselman, G. (2000). 3D reconstruction of building models. *International Archives of Photogrammetry and Remote Sensing*, *33*(B), 538–545.

Sýkora, L. (1994). Local urban restructuring as a mirror of globalization processes: Prague in the 1990s. *Urban Studies*, *31*(7), 1149–1166.

Tress, B., & Tress, G. (2003). Scenario visualisation for participatory landscape planning – a study from Denmark. *Landscape and Urban Planning*, *64*(3), 161–178.

Vaishar, A., et al. (2001). Small towns: An important part of the Moravian settlement system. In M. Pak & D. Rebernik (Eds.), *Cities in transition* (pp. 309–318). Ljubljana: Univerze v Ljubljani.

Walz, A., et al. (2007). Participatory scenario analysis for integrated regional modelling. *Landscape and Urban Planning*, *81*(1–2), pp. 114–131.

Williams, M. (1994). The relations of environmental history and historical geography. *Journal of Historical Geography*, *20*(1), 3–21.

Chapter 12
Does Landscape Structure Reveal Ecological Sustainability?

Christa Renetzeder, Thomas Wrbka, Sander Mücher, Michiel van Eupen, and Michiel Kiers

12.1 Concepts of Landscape Ecology for Sustainability Impact Assessment

In the last decades, sustainable development has been acknowledged to be of major importance in the European Union for the future, leading to numerous Commission Strategies and Action Plans (amongst others CEC, 2001, 2004, 2005a, 2005b, 2005c, 2007). Sustainability Impact Assessment is one option to convert policies into operational approaches. Considerable efforts are made to assess the impact of policy on the three pillars comprised of economic, social and environmental to ensure a holistic benefit for society. Indicators (e.g. EEA, 2005) and guidelines have been developed to enable an in-depth analysis of sustainability impact assessment (CEC, 2005d). Helming, Pérez-Soba, & Tabbush (2008) and Peterseil et al., (2004) showed that changes in a landscape regarding the anthropogenic influence are a highly integrative indicator for sustainability. The structure of the landscape reflects not only the natural settings of the landscape but also its history and the impact of mankind throughout the centuries (Antrop, 2005; Ernoult, Freire-Diaz, Langlois, & Alard, 2006), so the present spatial patterns are a result of former activities and processes in the landscape. This "pattern and process" paradigm is a key concept in modern landscape ecology (Turner, Gardner, & O'Neill, 2001). It relates to the spatial and functional relationship between distinct local ecosystems by describing the distribution of energy, resources and species in relation to size, shape, number and type of ecosystems in a particular landscape. Many European cultural landscapes developed their own regionally distinct pattern of landscape elements which has been the aim of ongoing or recently finalised European research projects. Human influence tends to result in a simplification and geometrisation of landscape pattern (Forman, 1995; Turner et al., 2001) and impact on ecological values (Millennium Ecosystem Assessment, 2003; Peterseil et al., 2004; Zebisch,

C. Renetzeder (✉)
Department of Conservation Biology, Vegetation & Landscape Ecology, University of Vienna, Rennweg 14, A-1030, Vienna, Austria
e-mail: christa.renetzeder@univie.ac.at

J. Anděl et al. (eds.), *Landscape Modelling*, Urban and Landscape Perspectives 8,
DOI 10.1007/978-90-481-3052-8_12,

Wechsung, & Kenneweg, 2004) and biodiversity (Moser et al., 2002; Pino, Rodà, Ribas, & Pons, 2000; Schindler, Kati, & Poirazidis, 2007; Zechmeister & Moser, 2001; Zechmeister et al., 2003) of landscapes have been shown. The removal of small biotopes or changes in the patch size of land-use parcels to larger units can therefore be seen as an unsustainable development, at least in terms of ecological sustainability. Opposite processes, such as the introduction of small biotopes, can be seen as sustainable development in its ecological dimension.

Complex biophysical, historical and political patterns in Europe lead to considerable regional differences in economic, social and environmental situations (Jongman, 2002). In general, the current knowledge about landscape structure and sustainability comes from case studies (Turner & Ruscher, 1988; Moser et al., 2002; Wrbka et al., 2004). Consequently, consistent and statistically robust information on structure in the different parts of Europe is still missing. To achieve an accurate spatially explicit sustainability impact assessment, the regional complexities need to be considered and indicator systems adapted to the regional level (Blaschke, 2006). Identifying and delineating spatial units that are relatively homogeneous in both biophysical and socio-economic contexts allow a successful upscaling of the approaches.

To meet these concerns, the Spatial Regional Reference Framework (Renetzeder, Van Eupen, Mücher, & Wrbka, 2008) was developed within the FP6-project SENSOR: In this integrated project, the main objective was to develop an ex-ante sustainability impact assessment tool including pan-European databases and spatial reference frameworks for the analysis of land and human resources in the context of European land-use policies. This approach was enabling the identification of European regions (EU 27 + Norway and Switzerland) that are to a certain extent similar in terms of their environmental, social and economic situations. Within these regions, thresholds and limits of European sustainability indicators were defined. The spatial unit is NUTS 2/3-regions, because many social and economic data are only available on the administrative level.

Herein, the process of a consistent European pattern analysis is examined, results of the differences in landscape structure among ten European regions are presented and the use for sustainability impact assessment of agricultural landscapes is discussed.

12.2 Material and Methods

Important input data for sampling design was the stratum of the Spatial Regional Reference Framework (SRRF – Renetzeder et al., 2008). It covers the EU 27, Norway and Switzerland and delineates 27 SRRF regions and identifies spatial administrative units on the basis of statistical clustering of biophysical (climate, topography and bedrock) and socio-economic data (including population density, population change rate, activity rate, gross domestic production, unemployment rate, functional urban areas and land cover data). This product was used in order to have a definition of more or less homogeneous regions and to randomly select sampling sites with the condition of representing European biogeographic regions (EEA, 2002). Selected SRRF-regions were (first biogeographic region, followed by main land-cover class): Boreal Forest (borf), Central Atlantic Mixed Agricultural Activities (catm), Continental Forest (conf), Continental Heterogeneous Agricultural Areas (conh), Mediterranean Arable Land (meda), Mountains Open Spaces (mnto), North Atlantic Arable Land (nata), North Atlantic Pastures (natp), Nemoral Mixed Agricultural Activities (nemi), Pannonian Arable Land (pana). Within the selected regions, three satellite images of $50 \times 50\ \text{km}^2$ are randomly selected using the official European grid system (Fig. 12.1).

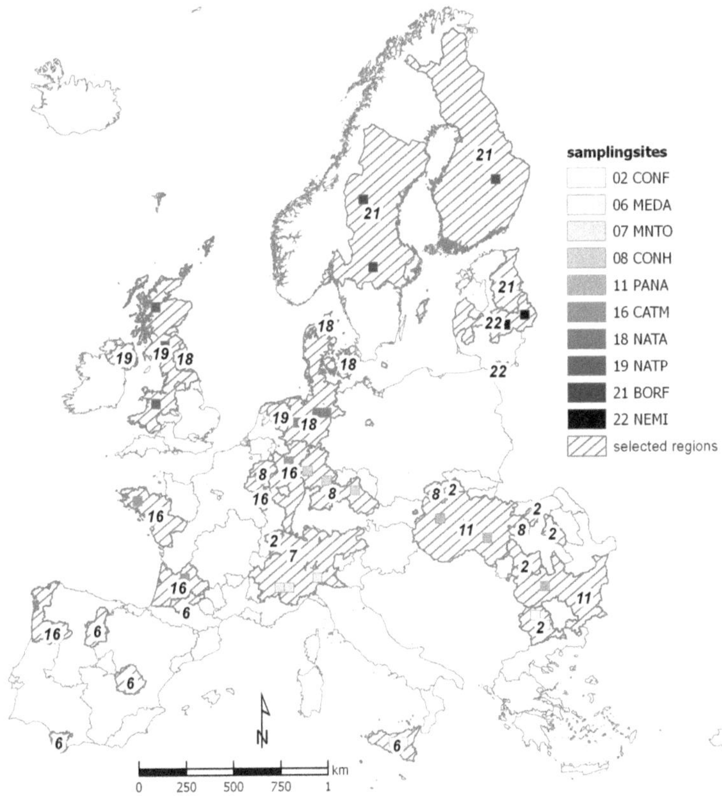

Fig. 12.1 Ten SRRF regions (Renetzeder et al., 2008) with three sampling sites (50×50 km)[1]

The satellite images (Image2000) cover most of the European countries. They are provided by the Joint Research Centre (JRC; http://image2000.jrc.it/) and are publicly available. They are the main source of information regarding landscape structure. Segmentation of satellite images was performed with the software package eCognition (© Definiens Inc.) delivering pure structural information. In order to fill this "empty" layer of structural information with a thematic one, Corine Land Cover (CLC2000; http://terrestrial.eionet.europa.eu/CLC2000) was used. These land-cover data have a minimum mappable unit of 25 ha and a minimum width of 100 m. There are 44 classes in five major groups distinguished. With the usual GIS methods, the structural information was intersected with CLC2000 in ArcGIS (ESRI, Inc., Redlands, CA) appointing a thematical meaning to the individual landscape element. Because of differences in the Minimum Mappable Unit (MMU) and minimum width (MMU segmentation – 0.25 ha, minimum width – 25 m; MMU CLC2000 – 25 ha, minimum width – 100 m) more detailed structural than thematic information is available. The combination of all input data allowed the building of a consistent database of landscape-structure data.

Calculation of the landscape-structure indices was done with V-Late 1.1, Vector-based Landscape Analysis Tools Extension (Lang & Tiede, 2003). V-Late is able to calculate landscape metrics on three different levels – the patch level (metrics of every single landscape element), the class level (averages of the metrics of all landscape elements belonging to one land cover) and the landscape level (average over all elements within one landscape no matter which thematic information they possess). We selected five landscape metrics: Patch Size (MPS_ha), Patch Edge (MPE), Shape Index (MSI), Perimeter-Area-Ratio (MPAR), and Fractal Dimension (MFRACT).

Focus was given to those land-cover classes which are most interesting in agricultural landscapes and which were present in at least nine selected regions with a high percentage in area: non-irrigated arable land, pastures, complex cultivation patterns, agricultural land with natural vegetation, and natural grasslands.

Statistical analysis was performed with R2.6.0 (R Development Core Team, 2006) and Statgraphics 5.0 with the main focus on extracting differences of landscape metrics regarding land-cover classes among the regions. Standard statistical procedures were applied testing for normal distribution and variance homogeneity. Mann–Whitney U test, discriminant and factor analysis were performed to detect differences in terms of class-level landscape metrics among the regions. In particular, the factor analysis with varimax rotation was carried out to determine the contribution of the individual landscape metrics to the variation in the data at class level (Cumming & Vernier, 2002; Schindler, Poirazidis, & Wrbka, 2008). Quadratic discriminant function analysis (QDA) was performed to test whether the function with the variables classified according to the existing groups, and thus whether the multivariate set of landscape metrics allows an accurate separation of the data into SRRF-regions.

12.3 Results

In the different regions, the area of land-cover classes varied enormously (Table 12.1). In all regions, agricultural and forest classes are the dominating land cover. The share of anthropogenic, seminatural and natural cover is rather low.

Significant differences of single landscape-structure indices among the regions were detected by box-and-whisker plots and the Mann–Whitney U test. Not every landscape metric depicted differences where others did, also some variation between the land-cover classes were exposed. Interestingly, the structure of natural grassland

Table 12.1 Percentages of land-cover classes (label and class number) in the investigated SRRF-regions

Label	Borf	Catm	Conf	Conh	Meda	Mnto	Nata	Natp	Nemi	Pana
Continuous urban fabric		0.1	0.1	0.1	0.5	0.8	<0.1			<0.1
Discontinuous urban fabric	0.8	6.8	5.2	7.0	0.2	7.9	3.7	0.3	1.0	5.9
Industrial or commercial units	0.2	0.8	1.1	0.9	0.1	2.1	0.2	<0.1	0.2	0.6
Road and rail networks and associated land		0.1	0.1	0.1	<0.1	0.1	<0.1	0.1	<0.1	<0.1
Port areas		<0.1		<0.1						<0.1
Airports	<0.1	0.2	0.1	0.1		0.1	0.1		<0.1	<0.1
Mineral extraction sites	<0.1	0.2	1.3	0.2	0.1	0.1	<0.1	<0.1	<0.1	<0.1
Dump sites		<0.1	0.2	<0.1		<0.1	<0.1			<0.1
Construction sites		0.1		<0.1		<0.1			<0.1	<0.1
Green urban areas	<0.1	0.3	0.2	0.2		0.3	<0.1		<0.1	<0.1
Sport and leisure facilities	<0.1	0.3	0.2	0.3	<0.1	0.2	0.1	0.2	<0.1	0.1
Non-irrigated arable land[a]	1.9	*19.7*	*20.1*	*38.6*	*54.6*	*34.8*	*48.4*	1.0	*24.4*	*74.9*
Permanently irrigated land					0.5					
Rice fields						3.3				0.1
Vineyards		1.8	1.5	0.7	0.6	0.3				1.5
Fruit trees and berry plantations		0.6	1.1	0.6	2.8	0.1			0.1	0.2
Olive groves					3.5					
Pastures[a]	<0.1	11.8	3.3	4.7		2.4	12.2	*21.6*	*14.3*	4.7
Annual crops with permanent crops					0.7					
Complex cultivation patterns[a]	0.2	*24.9*	4.0	9.6	4.5	4.2	4.5	1.2	12.0	1.8
Agricultural land with of natural vegetation[a]	2.1	2.4	8.2	1.6	2.7	3.0	3.6	2.2	9.6	1.7
Agro-forestry areas					<0.1	<0.1				
Broad-leaved forest	2.1	*14.7*	*31.2*	*13.0*	3.9	8.7	1.6	0.9	*13.3*	4.3
Coniferous forest	*43.7*	3.2	5.5	*16.3*	1.6	*10.4*	*20.1*	*13.8*	5.3	0.1
Mixed forest	*13.6*	*10.4*	6.5	4.5	2.5	1.1	2.1	<0.1	12.0	0.2
Natural grasslands[a]	0.1	<0.1	2.6	1.0	6.6	8.2	0.4	*20.2*		0.6
Moors and heathland					<0.1	<0.1				
Sclerophyllous vegetation					7.0					
Transitional woodland-shrub	<0.1	1.0	5.4	0.1	3.7	2.3	0.1	1.9	4.6	1.1

Table 12.1 (continued)

Label	Borf	Catm	Conf	Conh	Meda	Mnto	Nata	Natp	Nemi	Pana
Beaches, dunes, sands			0.1		<0.1	0.2				0.3
Bare rocks	<0.1		0.1		0.1	4.4		0.7		
Sparsely vegetated areas	0.1		0.3	<0.1	0.7	4.1	0.3	5.2		
Burnt areas						<0.1				
Glaciers and perpetual snow						<0.1				
Inland marshes	<0.1	<0.1	0.3	<0.1		<0.1	<0.1	<0.1	0.2	0.4
Peat bogs	5.0		0.2				0.7	0.8	0.7	
Water courses	*14.0*	0.6	1.0	0.5	0.9	0.3	0.1	0.1	2.1	1.6
Water bodies	2.0							1.6		

Explanation: numbers >10% are given in italics.
[a]Land-cover classes chosen for further analysis.

was least different among the regions and the landscape metrics for non-irrigated arable land show the most differences. The most contrasting regions were Continental Forests and Mountains Open Spaces. Nearly every metric in all land-cover classes differed significantly from each other. The most similar regions were Central Atlantic Mixed Agricultural Activities – North Atlantic Arable Land, Boreal Forest – Nemoral Mixed Agricultural Activities, Central Atlantic Mixed Agricultural Activities – Continental Forest and Mediterranean Arable Land – Nemoral Mixed Agricultural Activities (Fig. 12.2 as example).

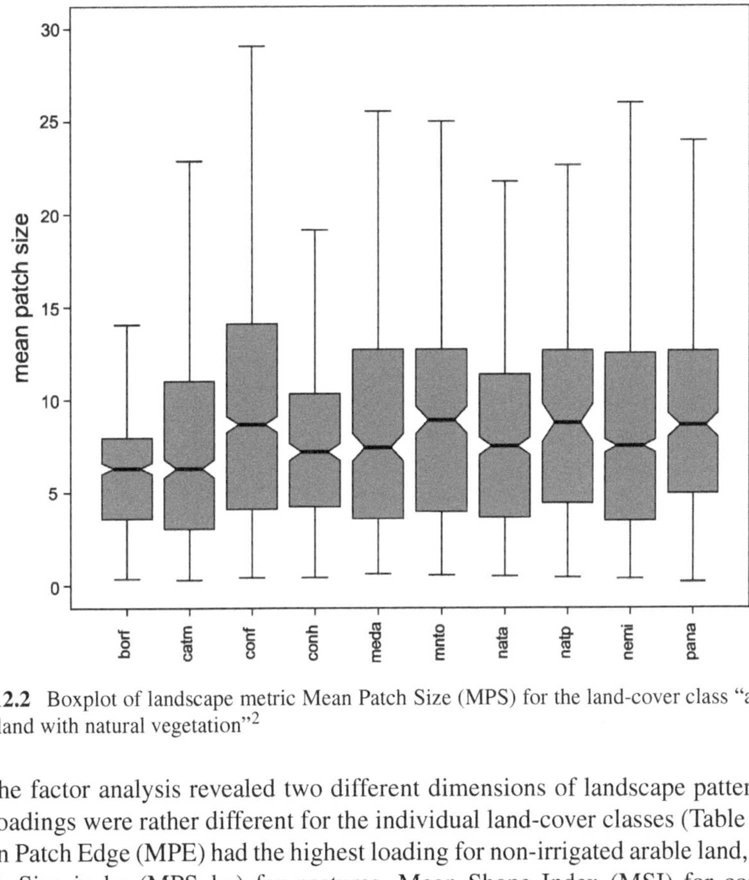

Fig. 12.2 Boxplot of landscape metric Mean Patch Size (MPS) for the land-cover class "agricultural land with natural vegetation"[2]

The factor analysis revealed two different dimensions of landscape pattern, but the loadings were rather different for the individual land-cover classes (Table 12.2). Mean Patch Edge (MPE) had the highest loading for non-irrigated arable land, Mean Patch Size in ha (MPS_ha) for pastures, Mean Shape Index (MSI) for complex cultivation pattern and agricultural land with natural vegetation and Mean Fractal Dimension (MFRACT) for natural grasslands. The QDA showed that the landscape-cover classes could only poorly be predicted by its combination of multivariate landscape metrics among the SRRF regions. In general, the error rate is rather high.

Spatial visualisation via Geographical Information Systems (GIS) gives a useful tool for the communication of statistical results. For example, the elements that match certain statistical results (above or below the Median etc.) can be marked accordingly (Fig. 12.3) and gives a quick overview over a certain area of interest.

Table 12.2 Factor Analysis at the European scale including the five class level metrics for each land-cover class[3]

	Non-irrigated arable land		Pastures		Complex cultivation patterns	
	Factor1	Factor2	Factor1	Factor2	Factor1	Factor2
MPS_ha	0.930	0.260	0.900	0.336	0.735	–0.630
MPE	0.997		0.822	0.565	0.879	–0.471
MSI	0.905	0.385	0.501	0.844	0.980	
MPAR	–0.380	0.760	–0.762	0.326	0.808	
MFRACT	0.515	0.854	–0.102	0.992	0.869	0.489
	Agricult. land with natural. veg.			Natural grasslands		
	Factor1		Factor2	Factor1		Factor2
MPS_ha	0.778		– 0.584	0.933		0.256
MPE	0.916		– 0.395	0.879		0.472
MSI	0.984			0.588		0.783
MPAR	0.808			–0.741		0.357
MFRACT	0.826		0.560	0.997		

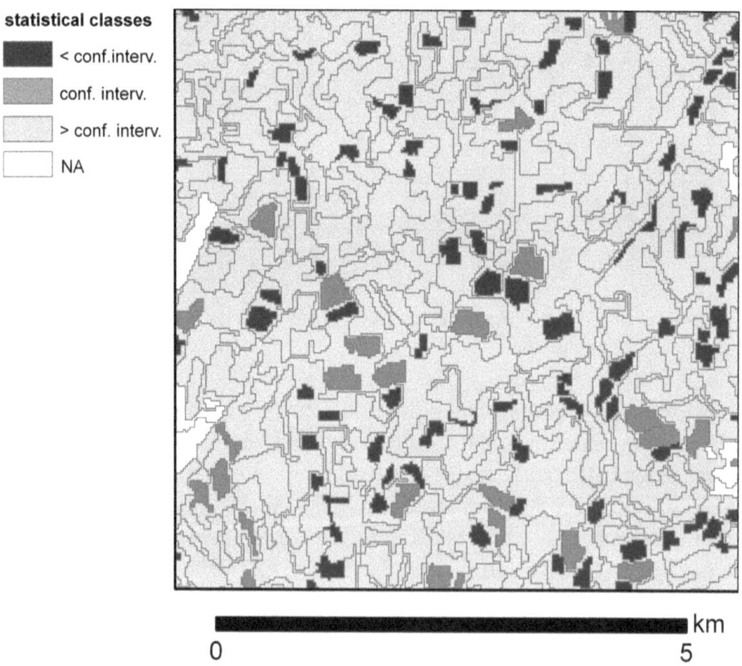

Fig. 12.3 Detail of the region "North Atlantic Arable Land" showing three statistical classes of the perimeter of land-cover class "Arable Land"[4]

12.4 Discussion and Conclusions

Various authors have criticised approaches (Estregueil et al., 2001, Haines-Young & Chopping, 1996; Kozak, Estreguil, & Ostapowicz, 2008) where land-cover information with relatively coarse spatial resolution is used for large-scale landscape studies. Corine land-cover classes reflect only the dominant land use but contain also other land-cover types and therefore, important structural features are not captured at this resolution. In contrast, high-resolution images and fine grain habitat and land-use maps are used to explain small-scale ecological phenomena on the landscape level (Lee, Ellis, Kweon, & Hong, 2008; Schindler et al., 2007, 2008). Consequently, large scale and widely used datasets such as Corine land cover should be supplemented and combined with an innovative derivation of homogeneous spatial units by segmentation as presented in our study.

Statistical analysis showed contradicting results concerning the differentiation of landscape metrics for the individual SRRF regions. Univariate analysis revealed significant differences between many regions for several land-cover classes (Fig. 12.2), multivariate analysis did not. Overlaying different distributions of the individual landscape metrics may have contributed to blurring of individual differences. The use of administrative units (NUTS 2/3 implemented into the SRRF) has weakened the desirable homogeneity in terms of primary and to some extent also secondary landscape structure, by resulting in lesser meaningful units for ecological assessment as confirmed by the poor performance of the set of metrics in the QDA. Further improvement of results may also be achieved by a higher number of sampling sites per region. Higher spatial resolutions could lead to a better discrimination of size and shape of patches. But at the same time higher resolutions will hamper the analysis for vast areas across Europe.

The use of landscape metrics as indices for Sustainability Impact Assessment is not very common (Graymore, Sipe, & Rickson, 2008) although certain aspects of ecological sustainability like biodiversity (Moser et al., 2002; Schindler et al., 2007) or naturalness (Peterseil et al., 2004) were investigated. Already Odum and Turner have shown in their classical work (1989) how increasing consumption of fossil energy and agrochemicals are coupled with a geometrical simplification of landscapes expressed by a decrease in fractal dimension. We have shown that even at the European scale, single landscape metrics react differently depending on land cover, suggesting that such sensitive indices could be utilised as indicators for ecological sustainability of land use, when applied together with land-cover types in a spatial regional reference. The elaboration of a scientifically sound knowledge-base for any impact assessment requires the establishment of statistically valid relationships between observed pattern and a particular ecological process of interest. This is especially true for biodiversity and sustainability related studies at the landscape level (Bunce et al., 2008) and calls for not only the combined use of geodatasets with different spatial resolution, but also for the inclusion of empirical data derived by representative field observations. As at European level, there is no consistent data available on human impact or hemeroby to correlate with, application of simple rules can be used for assessing trends in sustainability. Intensification of land use generally leads to simpler geometry and configuration of landscape elements (Forman 1995; Turner et al., 2001). A first indication is the deviation of landscape metrics from the average, based on the Austrian "Concept of Relative Deviance" (Peterseil et al., 2004) and currently applied on the presented European data.

Acknowledgements This study is part of the integrated project SENSOR (http://sensor-ip.org) financed via the 6th framework program of the European Commission. Our thanks are addressed to Katharina Helming as the project leader and Marta Perez-Soba as the module coordinator.

Notes

1. BORF = Boreal Forest, CATM = Central Atlantic Mixed agricultural activities, CONF = Continental Forest, CONH = Continental Heterogeneous agricultural areas, MEDA = Mediterranean Arable land, MNTO = Mountains Open spaces, NATA = North Atlantic Arable land, NATP = North Atlantic Pastures, NEMI = Nemoral Mixed agricultural activities, PANA = Pannonian Arable.
2. Overlapping notches (confidence interval of the Median) reveal regions which are not significantly different in this structure index.
3. Mean Patch Size in ha (MPS_ha), Mean Patch Edge (MPE), Mean Shape Index (MSI), Mean Perimeter-Area-Ratio (MPAR) and Mean Fractal Dimension (MFRACT).
4. Below and above the confidence interval of the Median, and the confidence interval.

References

Antrop, M. (2005). Why landscapes of the past are important for the future? *Landscape and Urban Planning*, *70*(1–2), 21–34.

Blaschke, T. (2006). The role of the spatial dimension within the framework of sustainable landscapes and natural capital. *Landscape and Urban Planning*, *75*(3–4), 198–226.

Bunce, R. G. H., Metzger M. J., Jongman R. H. G., Brandt J., De Blust G., Elena-Rossello R., et al. (2008). A standardized procedure for surveillance and monitoring European habitats and provision of spatial data. *Landscape Ecology*, *23*(1), 11–25.

CEC (Commission of the European Communities) (2001). *A Sustainable Europe for a Better World: A European Union Strategy for Sustainable Development. (Commission proposal to the Gothenburg European Council)*, COM(2001) 264, EC, Brussels.

CEC (2004). *Stimulating Technologies for Sustainable Development: An Environmental Technologies Action Plan for the European Union, Communication from the Commission to the Council, the European Parliament, the European Economic and Social Committee and the Committee of the Regions*, COM(2004) 38, EC, Brussels.

CEC (2005b). *The 2005 Review of the EU Sustainable Development Strategy: Initial Stocktaking and Future Orientations*, COM(2005) 37, EC, Brussels.

CEC (2005a). *Review of the Sustainable Development Strategy – A platform for action*, COM(2005) 658, EC, Brussels.

CEC (2005c). *Thematic Strategy on the sustainable use of natural resources*, COM(2005) 670 final, EC, Brussels.

CEC (2005d). *Impact assessment guidelines*, SEC(2005) 791, EC, Brussels.

CEC (2007). *Progress Report on the Sustainable Development Strategy 2007*, COM(2007) 642 final, EC, Brussels.

CLC (2000). *Corine Land Cover*, http://terrestrial.eionet.europa.eu/CLC2000

Cumming, S., & Vernier, P. (2002). Statistical models of landscape pattern metrics, with application to regional scale dynamic forest simulations. *Landscape Ecology*, *17*(5), 433–444.

EEA (European Environmental Agency) (2002). *Europeś biodiversity – bio geographical regions and seas*, EC, Copenhagen, http://reports.eea.eu.int/report_2002_0524_154909/en/tab_content_RLR

EEA (European Environmental Agency) (2005). *Agriculture and environment, in EU-15 – the IRENA indicator report*, EEA Report 6, 2005, EC, Copenhagen.

Ernoult, A., Freire-Diaz, S., Langlois, E., & Alard, D. (2006). Are similar landscapes the result of similar histories? *Landscape Ecology*, *21*(5), 631–639.

Estreguil, C., et al. (2001). *Relevance of corine land cover for reporting on nature protection at regional level*. Ispra, Italy: DG-JRC.

Forman, R. T. T. (1995). *Land mosaics – The ecology of landscapes and regions*. Cambridge: University Press.

Graymore, L. M. M., Sipe, N. G., & Rickson, R. E. (2008). Regional sustainability: How useful are current tools of sustainability assessment at the regional scale? *Ecological Economics*, *67*(3), 362–372.

Haines-Young, R., & Chopping, M. (1996). Quantifying landscape structure: A review of landscape indices and their application in forested landscapes. *Progress in Physical Geography*, *20*(4), 418–445.

Helming K., Pérez-Soba M., & Tabbush P. (Eds.). (2008). *Sustainability impact assessment of land use change*. Berlin, Heidelberg, New York: Springer.

JRC (Joint Research Centre, EC) *Image 2000 & Corine Land Cover 2000 Project*, http://image2000.jrc.it (22.10.2008)

Jongman, R. H. G. (2002). Homogenisation and fragmentation of the European landscape: Ecological consequences and solutions. *Landscape and Urban Planning*, *58*(2–4), 211–221.

Kozak, J., Estreguil, C., & Ostapowicz, K. (2008). European Forest cover mapping with high resolution satellite data: The Carpathians case study. *International Journal of Applied Earth Observation and Geoinformation*, *10*(1), 44–55.

Lang, S., & Tiede, D. (2003). *vLATE Extension für ArcGIS – vektorbasiertes Tool zur quantitativen Landschaftsstrukturanalyse*, ESRI Anwenderkonferenz 2003 Innsbruck. CD-ROM.

Lee, S. W., Ellis, C. D., Kweon, B. S., & Hong, S. K. (2008). Relationship between landscape structure and neighbourhood satisfaction in urbanized areas. *Landscape and Urban Planning*, *85*(1), 60–70.

Millennium Ecosystem Assessment (2003). *Ecosystems and human well-being: A framework for assessment*, Island Press, Washington.

Moser, D., Zechmeister, H., Plutzar, C., Sauberer, N., Wrbka, T., & Grabherr, G. (2002). Landscape patch shape complexity as an effective measure for plant species richness in rural landscapes. *Landscape Ecology*, *17*(7), 657–669.

Odum, E. P., & Turner, M. G. (1989). Georgia landscape: A changing resource. In I. S. Zonneveld & R. T. T. Forman (Eds.), *Changing landscapes: An ecological perspective* (pp. 137–146). Berlin, Heidelberg, New York: Springer.

Peterseil, J., Wrbka, T., Plutzar, C., Schmitzberger, I., Kiss, A., Szerencsits, E., et al. (2004). Evaluating the ecological sustainability of Austrian agricultural landscapes: The SINUS approach. *Land Use Policy*, *21*(3), 307–320.

Pino, J., Rodà, F., Ribas, J., & Pons, X. (2000). Landscape structure and bird species richness: Implications for conservation in rural areas between natural parks. *Landscape and Urban Planning*, *49*(1–2), 35–48.

R Development Core Team (2006) *R: A Language and Environment for Statistical Computing*, R Foundation for Statistical Computing, Vienna.

Renetzeder, C., Van Eupen M., Mücher C.A., & Wrbka T. (2008). A spatial regional reference framework for sustainability assessment. In K. Helming et al. (Eds.), *Sustainability impact assessment of land use changes* (pp. 249–268). Berlin, Heidelberg, New York: Springer.

Schindler, S., Kati, V., & Poirazidis, K. (2007). Testing the performance of landscape structure variables as predictors of biodiversity: A case study from Dadia NP, Greece. In R. G. H. Bunce et al. (Eds.), *25 years of landscape ecology: Scientific principles in practice – Proceedings of the 7th IALE world congress* (pp. 337–338). The Netherlands: Wageningen.

Schindler, S., Poirazidis, K., & Wrbka, T. (2008). Towards a core set of landscape metrics for biodiversity assessments: A case study from Dadia National Park, Greece. *Ecological Indicators, 8*(5), 502–514.

Turner M. G., Gardner R. H., & O'Neill R. V. (Eds.). (2001). *Landscape ecology in theory and practice, pattern and process*. Berlin, Heidelberg, New York: Springer.

Turner, M. G., & Ruscher, C. L. (1988). Changes in the spatial patterns of land use in Georgia, USA. *Landscape Ecology, 1*(4), 21–30.

Wrbka, T., Erb, K. H., Schulz, N. B., Peterseil, J., Hahn, C. O., & Haberl, H. (2004). Linking pattern and process in cultural landscapes. An empirical study based on spatially explicit indicators. *Land Use Policy, 21*(3), 289–306.

Zebisch, M., Wechsung, F., & Kenneweg, H. (2004). Landscape response functions for biodiversity – Assessing the impact of land use changes at the county level. *Landscape and Urban Planning, 67*(1–4), 157–172.

Zechmeister, H. G., et al. (2003). The influence of land-use practices and economics on plant species richness in meadows. *Biological Conservation, 114*(2), 165–177.

Zechmeister, H. G., & Moser, D. (2001). The influence of agricultural land-use intensity on bryophyte species richness. *Biodiversity and Conservation, 10*(10), 1609–1625.

Chapter 13
Landscape Approaches and GIS for Biodiversity Management

Stefan Schindler, Kostas Poirazidis, Aristotelis Papageorgiou, Dionisios Kalivas, Henrik Von Wehrden, and Vassiliki Kati

13.1 Landscape Approaches for Biodiversity Management

Landscape approaches and geographical information systems (GIS) have been playing an increasing role in biogeography and conservation biology over the last decade (Gaston, 2000; Foody, 2008; Gillespie, Foody, Rocchini, Giorgi, & Saatchi, 2008). Within this period, the number of papers using GIS published in the journal Landscape Ecology has roughly doubled (Anderson, 2008). Especially remote-sensing applications have grown in importance within recent years. Remote sensing now routinely provides environmental information ranging from global to local scales, and geographical information systems provide, among other applications, necessary interfaces to store, analyse and visualise spatial data; increased computational capacities triggered even more such applications. In this chapter, we demonstrate how the combination of landscape approaches, remote sensing and GIS aids conservation and management of biodiversity. We therefore summarise six case studies from Dadia National Park (Dadia NP), in north-eastern Greece. The studies aimed at (1) modelling of nesting habitat for a flagship species, (2) evaluation of land-use change, (3) detecting statistical dimensions and spatial patterns of landscape structure, (4) testing the performance of landscape metrics as indicators of biodiversity, (5) developing a GIS approach for a systematic raptor monitoring, and (6) developing a decision-support system to optimise conservation of biodiversity in managed forests.

13.2 Study Area and GIS Data

The study area, the Dadia NP, is situated in the Evros prefecture in north-eastern Greece (Fig. 13.1). Its extent of about 430 km^2 includes two strictly protected core areas covering 73.5 km^2. The mountainous area (altitudes ranging from 20 to 645 m above sea level) is covered by extensive pine (*Pinus brutia, P. nigra*)

S. Schindler (✉)
Department of Conservation Biology, Vegetation & Landscape Ecology, University of Vienna, Rennweg 14, A-1030, Vienna, Austria
e-mail: stefan.schindler@univie.ac.at

J. Anděl et al. (eds.), *Landscape Modelling*, Urban and Landscape Perspectives 8,
DOI 10.1007/978-90-481-3052-8_13, © Springer Science+Business Media B.V. 2010

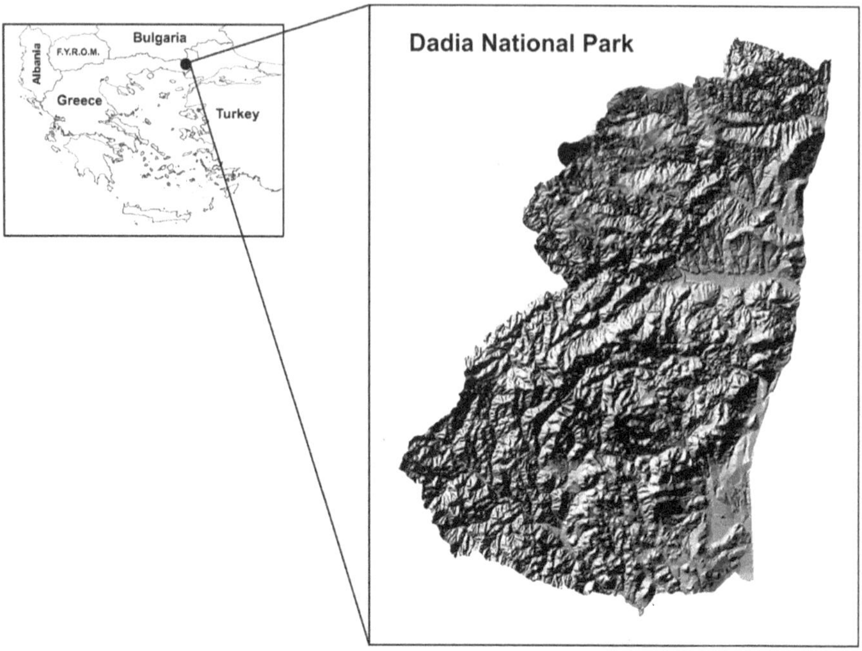

Fig. 13.1 Map of the case study area, Dadia National Park, located in NE Greece

and oak (*Quercus frainetto, Quercus cerris, Quercus pubescens*) forest, but it also includes a variety of other habitats such as pastures, agricultural fields, torrents and stony hills (Catsadorakis & Källander, 2010). Dadia NP is an essential refuge for breeding populations of a unique assemblage of raptors (Poirazidis et al., 1996, 2010a). It contains the only remaining Black Vulture (*Aegypius monachus*) breeding colony in the Balkan Peninsula (Poirazidis, Goutner, Skartsi, & Stamou, 2004; Skartsi, Elorriga, Vasilakis, & Poirazidis, 2008), and a high diversity of passerines (Kati & Sekercioglu, 2006), amphibians and reptiles (Kati, Fofopoulos, Ioannidis, Poirazidis, & Lebrun, 2007), butterflies (Grill & Cleary, 2003), grasshoppers (Kati, Dufrêne, Legakis, Grill, & Lebrun, 2004b), and vascular plants (Kati, Lebrun, Devillers, & Papaioannou, 2000; Korakis et al., 2006).

Satellite images (IKONOS, July 2001, pixel size 1 m) of the study area were digitised to produce a vector map including 14 different habitat types related to the dominant forest tree species and six classes of the percentage of mixed forest. Out of this initial habitat base map, further maps differing in the number of land-cover categories were produced for the case studies.

13.3 Case Study 1 – Modelling Nesting Habitat as a Conservation Tool for the Eurasian Black Vulture

This study[1] formulated habitat models in order to predict the potential nesting habitat of Black Vulture in Dadia NP, a priority breeding species for the area as well as over the rest of the Balkan Peninsula (Skartsi et al., 2008). The aims of this study were (1) to identify crucial determinants of suitable nesting habitat characteristics and (2) to build empirical models for the prediction of nesting habitat. Using logistic regression and 16 environmental variables, separate models regarding geomorphology, vegetation types, and disturbance factors were obtained and combined using Bayesian statistics. At the final stage a Boolean map of the mature forest refined the present suitable nesting habitat (Fig. 13.2). The geomorphology contributed more than all other predictors to the final overall model of a suitable Black Vulture nesting habitat. The nesting preference in areas with steep slopes seems to be adaptive, as such areas provide better foraging opportunities and protection from predators

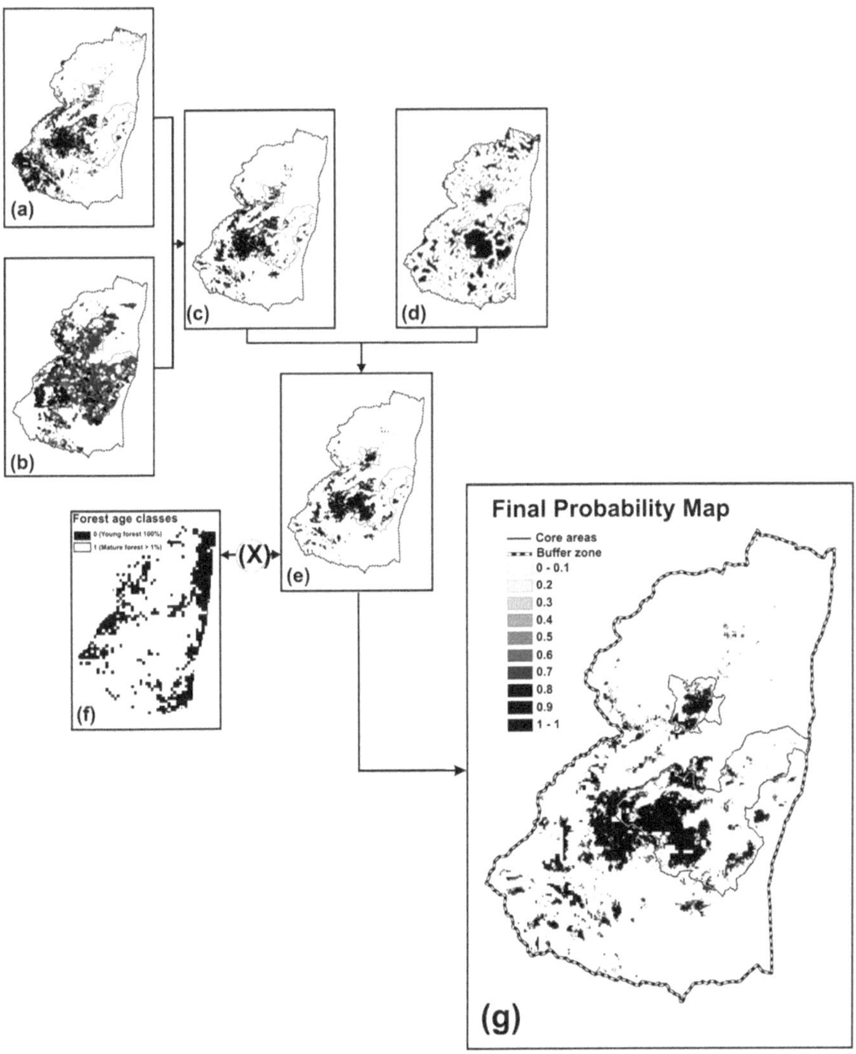

Fig. 13.2 Maps of probability of occurrence for the nest sites of the black vulture[3]

(Hiraldo & Donázar, 1990; Fargallo, Blanco, & Soto-Largo, 1998; Donázar, Blanco, Hiraldo, Soto-Largo, & Oria, 2002). The results of this study were used to improve Black Vulture Monitoring, forest management and the zonation of the National Park.

13.4 Case Study 2 – Forest Re-growth Since 1945 in the Dadia Forest Nature Reserve

In this study,[2] the focus was drawn on the interpretation of aerial photographs and satellite images in order to identify land-use patterns in Dadia NP for 1945, 1973 and 2001, and thus to quantify the land-use changes among these years. The landscape was classified to the three categories forest, openings, and agricultural land, and the most obvious change was a dramatic decline in forest openings (Table 13.1), caused mainly by land abandonment and reforestation programs. During a period of 50 years, the landscape lost part of its characteristic heterogeneity and mosaic-structured character, landscape qualities that are very important for the maintenance of biodiversity of several groups of organisms (Atauri & De Lucio, 2001; Torras, Gil-Tena, & Saura, 2008).

Table 13.1 Land-use change in Dadia National Park from 1945 to 2001

Land use	Zone	1945	1945–1973	1973	1973–2001	2001
		[km^2]	[%]	[km^2]	[%]	[km^2]
Forest	Core area	37.7	+33	50.1	+20	60.2
	Buffer zone	160.5	+15	183.9	+37	251.2
	Total area	198.2	+18	234.0	+33	312.6
Openings	Core area	33.3	–40	20.1	–50	10.1
	Buffer zone	119.4	–27	87.0	–67	28.6
	Total area	152.7	–30	107.1	–64	38.7
Agricultural land	Core area	1.9	+43	2.7	–40	1.6
	Buffer zone	76.4	+12	85.4	–21	67.2
	Total area	78.3	+13	88.1	–22	69.0

13.5 Case Study 3 – Towards a Core Set of Landscape Metrics for Biodiversity Assessments: A Case Study from Dadia National Park

Landscape metrics in the GIS environment can be used to facilitate the investigation of the relation between landscape structure and biodiversity (Hill & Curran, 2003; Honnay, Piessense, & Landuy, 2003). Data reduction analyses have been applied to tackle the problem of highly correlated indices (Riitters, Neill, & Hunsaker, 1995; Cushman, McGarigal, & Vell, 2008), but valid landscape predictors for fine-scale Mediterranean forest-mosaics have been missing. In this study,[4] we used a wide array of related variables of landscape structure, (1) to investigate correlations and statistical dimensions of landscape structure at landscape and class level, (2) to provide a core set of representative variables, (3) to evaluate the stability of the detected dimensions across scales, and (4) to describe the characteristic landscape pattern of Dadia NP. Therefore, we produced a map of nine land-cover categories that we converted to raster format with a grain of 5 m. We used FRAGSTATS (McGarigal & Marks, 1995) for the computation of the 119 landscape metrics investigated in the study and applied correlation analysis and factor analysis, regarding both landscape and class level metrics in a parallel way. Landscape diversity, edge contrast (a measure related to fragmentation) and area-weighted mean patch shape were stable at landscape level across the three tested scales. The representative set of metrics consisted of Simpson's Diversity Index, Mean Edge Contrast Index, and the Area-Weighted Mean Shape Index. The pattern analysis revealed a dispersed pattern for landscape diversity, with high values in the vicinity of the borders between core areas and the buffer zone, and a clustered pattern for edge contrast, presenting a gradient from the unfragmented core areas to the agricultural land in the east of the reserve (Fig. 13.3).

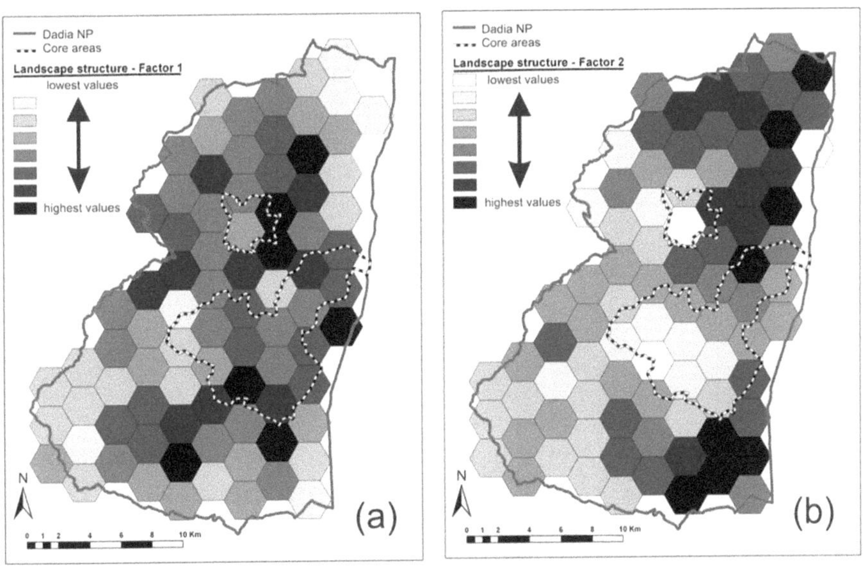

Fig. 13.3 Pattern of the main dimensions of landscape structure in Dadia National Park. (**a**) Landscape diversity (Factor 1) and (**b**) edge contrast (Factor 2)

13.6 Case Study 4 – Testing the Performance of Landscape Metrics as Indicators for Biodiversity

Since only some empirical studies tested the relations between landscape structure and the species diversity of multiple taxa (Hernández-Stefanoni, 2006; Yamaura,

Amano, & Katoh, 2008), we tried to fill this gap in this case study.[5] We analysed the relations of 52 landscape structure variables with overall biodiversity and with species richness of the six taxa woody plants, orchids, orthopterans, amphibians, reptiles and birds. Species data were collected by Kati et al. (2004a), based on standard methods; landscape structure variables were computed for circular areas of five different extents around the sampling plots. For each taxon the species richness was modelled with each individual landscape variable at each scale as the predictor, based on a linear mixed model using the software R (R Development Core Team, 2008). Additionally, we tested the performance of sets of three landscape-structure variables as predictors of species richness, using AIC to compare sets composed by different methods such as expert knowledge, several methods of ordination (see previous case study or Schindler, Poirazidis, & Wrbka, 2008), decision trees, random choice, and optimal sets after testing all possible combinations.

In this study, landscape metrics proved to be good indicators of species richness regarding the taxa woody plants, orthopterans, reptiles and for overall biodiversity. Metrics regarding patch shape, proximity, texture and diversity resulted frequently in significant univariate models, while metrics regarding similarity or edge contrast hardly contributed to significant models. Our results revealed that the scale affected the performance of landscape metrics. Woody plants, orthopterans and birds were better predicted at smaller scales, while reptiles were predicted best at larger scales. Regarding the different methods of composing sets, optimal sets performed always significantly better than all other methods. The statistical methods performed slightly better than random choice, while the expert knowledge performed slightly worse than random. The revealed pattern of relations and performances will be useful to understand landscape structure as driver and indicator of biodiversity, and to improve management decisions in Mediterranean forests and similar mosaic landscapes.

13.7 Case Study 5 – Development of a Geographic Information System for Territory Analysis of Raptor Species

Dadia National Park is well known for its high diversity of breeding birds of prey, a community in total exceeding 300 territories (Poirazidis et al., 2010a). An integrated monitoring plan was implemented by WWF – Greece in 1999, aiming at the effective conservation of biodiversity and ecological values of the area. In this case study[6] we describe the development of a GIS approach to estimate the territories of breeding raptors. All raptors within 34 permanent plots were counted and each plot was censused five times during the breeding seasons 2001–2005. Raptor observations were labelled in GIS, showing flight trajectories, possible nest sites, the number of synchronously observed individuals, age, sex, and different territorial activities under different symbols to enable analyses that consider all the information obtained in the field. The progressive analysis per species was based on eight criteria related to territorial behaviour, general observations and biology of the species as well as to landscape features (Poirazidis et al., 2006, 2010c). Breeding territories were differently classified as confirmed or possible. The GIS approach for estimating raptor territories was particularly effective for strictly territorial species like most of the eagles, buzzards, hawks, and falcons (Table 13.2). Less territorial species, such as the Egyptian Vulture (*Neophron percnopterus*) and the Short-toed Eagle (*Circaetus gallicus*) demanded a large amount of data to enable precise territory estimations.

Table 13.2 Summary of the species-specific problems and advantages of the GIS-based methodology for the estimation of raptor population sizes at local scale (values are scaled from 1 = not any to 6 = very high)

Species	Problems with low territoriality	Problems with secretiveness or late arrival	Frequent key observations, high accuracy	Total usefulness GIS method
White-tailed Eagle	4	2	4	4
Golden Eagle	1	2	4	6
Imperial Eagle	2	5	4	4
Lesser spotted Eagle	3	2	5	6
Short-toed Eagle	5	1	4	4
Booted Eagle	2	3	4	5
Egyptian Vulture	6	3	6	5
Common Buzzard	2	1	5	6
Long-legged Buzzard	2	3	5	6
Honey Buzzard	2	5	4	4
Black Kite	6	3	2	3
Marsh Harrier	5	2	2	3
Goshawk	1	4	3	4
Levant Sparrowhawk	2	6	3	3
Sparrowhawk	2	4	3	4
Peregrine Falcon	3	2	5	6
Lanner Falcon	2	2	5	6
Hobby	1	5	3	4
Eurasian Kestrel	1	2	4	6
Black Stork	6	1	4	4

13.8 Case Study 6 – Conservation of Biodiversity in Managed Forests: An Integrated Approach Using Multi-Function Forest Services

In this case study[7] we developed a decision-support system to optimise the conservation of biodiversity in managed forests. We investigated timber production and biodiversity, the main ecosystem services of the Mediterranean forest landscape of Dadia NP. We produced (1) a series of spatially explicit habitat suitability models for higher plants, amphibians, small forest birds and raptors and an overall model for total local biodiversity, (2) maps related to timber production and (3) three management scenarios and a decision-support system based on a conflict assessment. Thus, we were able to establish integrated management concepts, and to assess the effects of different management strategies on the two main ecosystem services.

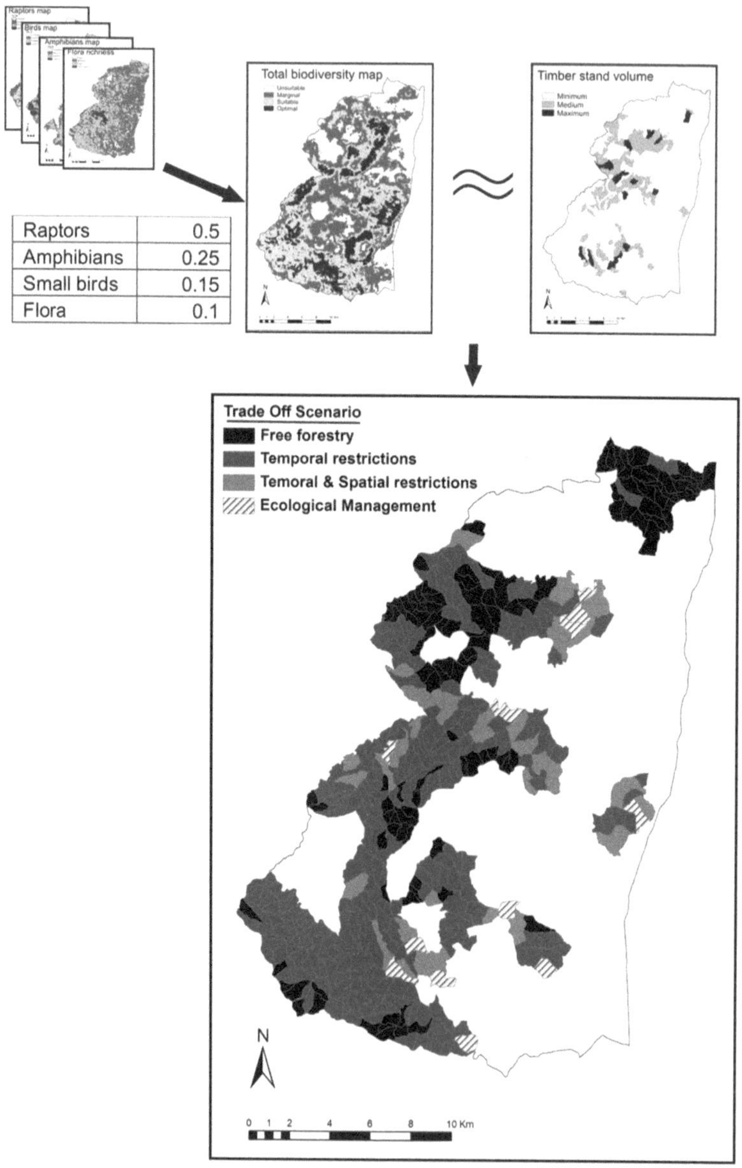

Fig. 13.4 Map of Dadia NP after the trade-off scenario considering conservation of biodiversity and timber production

Spatial modelling was based on data of several systematic field surveys. We used 23 eco-geographical variables to derive predictors for species habitat suitability, and modelled five taxa as surrogates for the total biodiversity in Dadia NP, namely grasses and shrubs (combined later to "higher plants"), amphibians, small forest birds (mainly Passerines) and raptors. For the three groups of fauna we created species distribution maps, while regarding plant species we used the accumulated number of plant species as a proxy of biodiversity. For the raptor data set (Poirazidis et al., 2006) we pooled data from 5 years and plotted the centre of their yearly territories. All the data were converted to a raster grain of 50 × 50 m, and Environmental Niche Factor Analysis (ENFA) was performed within the BIOMAPPER software (version 3.2; Hirzel, Hausser, Chessel, & Perrin, 2002). The total timber standing volume per sub-section was estimated using the official forest service inventory for the current forest management plan 2006–2016. The relative thematic maps were classified into four bins, (1) unsuitable, (2) marginal, (3) suitable and (4) optimal regarding habitat suitability, and (1) minimum, (2) medium, (3) large and (4) maximum regarding timber stand volume. We considered four different forest-management actions at the stand level: management (1) without limitations, (2) with temporal restrictions, (3) with temporal and spatial restrictions and (4) focused on the ecological values. Three general management scenarios were formulated: Conservation, timber production and trade off. A major output was the map of the proposed forest-management categories of the trade-off scenario (Fig. 13.4).

13.9 Conclusions and Implications for Biodiversity Management

Landscape approaches involving GIS and integrated statistical approaches proved to be useful to understand the relations of pattern and changes of landscape structure with the present biodiversity and the habitat suitability for different groups of organisms. This knowledge was essential to establish conservation strategies for biodiversity, for instance regarding the maintenance of habitat heterogeneity in both the core and buffer zone of the reserve (Grill & Cleary, 2003; Kati et al., 2004b; Kati and Sekercioglu, 2006), and for the optimisation of other ecosystem services such as timber production. Habitat suitability modelling for selected groups of organisms to develop management scenarios for managed forests is highly recommendable.

Landscape surveillance should be integrated into the ecological monitoring of key and indicator species to aid the evaluation of the management effects on both forest and wildlife. Further research regarding species, taxa and landscape indicators on a larger scale would be desirable to further extrapolate and validate the models, and enable an even more complete strategy for biodiversity conservation and management.

Acknowledgements We are very grateful to the colleagues and volunteers from WWF Greece/Dadia project who collaborated in the case studies described herein. We thank Christa Renetzeder for her helpful comments on a previous version of the manuscript.

Notes

1. This and the following case studies draw upon the already published papers. Thus, for each study we indicate the concrete reference. Study 1 emerges from Poirazidis et al. (2004).
2. Triantakonstantis et al. (2006).
3. (a) A geomorphological model, (b) a vegetation-type model, (c) a model combining a and b, (d) a disturbance model, (e) a model combining c and d, (f) a Boolean map of mature forest, and (g) the final map combining e and f.
4. Schindler et al. (2008).
5. Schindler et al. (2009).
6. Poirazidis et al. (2006, 2009).
7. Poirazidis et al. (2008, 2010b).
8. The managed forests are categorised into the four management options free forestry, temporal restrictions, temporal and spatial restrictions, and ecological management.

References

Anderson, B. J. (2008). Research in the journal landscape ecology, 1987–2005. *Landscape Ecology*, *23*(2), 129–134.

Atauri, J. A., & De Lucio, J. V. (2001). The role of landscape structure in species richness distribution of birds, amphibians, reptiles and lepidopterans in Mediterranean landscapes. *Landscape Ecology*, *16*(2), 147–159.

Catsadorakis, G., & Källander, H. (2010). *The Dadia – Lefkimi – Soufli forest national park.* Greece: Biodiversity, Management and Conservation. WWF Greece, Athens, Greece (in press).

Cushman, S. A., McGarigal, K., & Neel, M. C. (2008). Parsimony in landscape metrics: Strength, universality, and consistency. *Ecological Indicators*, *8*(5), 691–703.

Donázar, J. A., Blanco, G., Hiraldo, F., Soto-Largo, E., & Oria, J. (2002). Effects of forestry and other land-use practices on the conservation of Cinereous Vultures. *Ecological Applications*, *12*(5), 1445–1456.

Fargallo, J. A., Blanco, G., & Soto-Largo, E. (1998). Forest management effects on nesting habitat selected by Eurasian Black Vultures Aegypius monachus in central Spain. *Journal of Raptor Research*, *32*(3), 202–207.

Foody, G. M. (2008). GIS: Biodiversity applications. *Progress in Physical Geography*, *32*(2), 223–235.

Gaston, K. J. (2000). Global patterns in biodiversity. *Nature*, *405*, 220–227.

Gillespie, T. W., Foody, G. M., Rocchini, D., Giorgi, A. P., & Saatchi, S. (2008). Measuring and modeling biodiversity from space. *Progress in Physical Geography*, *32*(2), 203–221.

Grill, A., & Cleary, D. F. R. (2003). Diversity pattern in butterfly communities of the Greek nature reserve Dadia. *Biological Conservervation*, *114*(3), 427–436.

Hernández-Stefanoni, J. L. (2006). The role of landscape patterns of habitat types on plant species diversity of a tropical forest in México. *Biodiversity and Conservation*, *15*(4), 1441–1458.

Hill, J. L., & Curran, P. J. (2003). Area, shape and isolation of tropical forest fragments. Effects on tree species diversity and implications for conservation. *Journal of Biogeography*, *30*(9), 1391–1403.

Hiraldo, F., & Donázar, J. A. (1990). Foraging time in the Cinereous Vulture Aegypius monachus: Seasonal and local variations and influence of weather. *Bird Study*, *37*, 128–132.

Hirzel, A. H., Hausser, J., Chessel, D., & Perrin, N. (2002). Ecological-niche factor analysis: How to compute habitat suitability maps without absence data?. *Ecology*, *83*(7), 2027–2036.

Honnay, O., Piessens, K., & Landuyt, W. V. (2003). Satellite based land use and landscape complexity indices as predictors for regional plant species diversity. *Landscape and Urban Planning*, *63*(4), 241–250.

Kati, V., Lebrun, P., Devillers, P., & Papaioannou, H. (2000). Les orchidées de la réserve de Dadia (Grèce), leurs habitats et leur conservation. *Les Naturalistes Belges, 81*, 269–282.

Kati, V., Devillers, P., Dufrene, M., Legakis, A., Vokou, D., & Lebrun, P. (2004a). Testing the value of six taxonomic groups as biodiversity indicators at a local scale. *Conservation Biology, 18*(3), 667–675.

Kati, V., Dufrêne, M., Legakis, A., Grill, A., & Lebrun, P. (2004b). Conservation management for Orthoptera in the Dadia reserve, Greece. *Biological Conservation, 115*(1), 33–44.

Kati, V., Fofopoulos, J., Ioannidis, Y. H. P., Poirazidis, K., & Lebrun, P. (2007). Diversity, ecological structure and conservation of herpetofauna in a Mediterranean area (Dadia National Park, Greece). *Amphibia-Reptilia, 28*(4), 517–529.

Kati, V., & Sekercioglu, C. H. (2006). Diversity, ecological structure, and conservation of the landbird community of Dadia reserve, Greece. *Diversity and Distribution, 12*(5), 620–629.

Korakis, G., et al. (2006). Floristic records from Dadia-Lefkimi-Soufli National Park, NE Greece. *Flora Mediterranea, 16*, 11–32.

McGarigal, K., & Marks, B. (1995). *FRAGSTATS: Spatial pattern analysis program for quantifying landscape structure*, USDA Forest Service General Technical Report PNW-GTR-351, Pacific Northwest Research Station, Portland.

Poirazidis, K., Skartsi, T., Pistolas, K., & Babakas, P. (1996). Nesting habitat of raptors in Dadia reserve, NE Greece. In J. Muntaner & J. Mayol (Eds.), *Biología y Conservación de las Rapaces Mediterraneas*, 1994 (Vol. 4, pp. 325–333), Monografías, SEO, Madrid.

Poirazidis, K., et al. (2006). Development of a geographic information system for territory analysis of raptor species (pp. 15–15), *Proceedings of the 21st European conference for ESRI users*, ESRI, Marathon, Athens.

Poirazidis, K., et al. (2008). Conservation of biodiversity in managed forests: An integrated approach using multi-function forest services. In J. Chen, S., Liu, R. Lucas, P. Sun, R. Lafortezza, & L. Delp (Eds.), *Proceedings of the international conference landscape ecology and forest management: Challenges and solutions* (pp. 142–143). Chengdu: IUFRO Landscape Ecology.

Poirazidis, K., et al. (2010a). Diurnal birds of prey in Dadia-Lefkimi-Soufli National Park: Long-term population trends and habitat preferences. In: G. Catsadorakis, & H. Källander (Eds.), *The Dadia-Lefkimi-Soufli Forest National Park, Greece: Biodiversity, management and conservation* (pp.151–168). Greece: WWF, Athens.

Poirazidis, K., et al. (2010b). Conservation of biodiversity in managed forests: Developing an adaptive decision support system. In C. Li, R. Lafortezza, & J. Chen (Eds.), *Landscape ecology in forest management and conservation. Challenges and solutions for global change* (pp. 103–114). Berlin, Heidelberg, New York: Higher Education Press – Springer.

Poirazidis, K., et al. (2009). *Monitoring breeding raptor populations – a proposed methodology using repeatable methods and GIS*. Avocetta *33* (in press).

Poirazidis, K., Goutner, V., Skartsi, T., & Stamou, G. (2004). Modelling nesting habitat as a conservation tool for the Eurasian black vulture (Aegypius monachus) in Dadia Nature Reserve, northeastern Greece. *Biological Conservation, 118*(2), 235–248.

R Development Core Team (2008). *R: A Language and Environment for Statistical Computing. R Foundation for Statistical Computing*, Vienna, Austria, http://www.R-project.org

Riitters, K. H., Neill, R. V., & Hunsaker, C. T. (1995). A factor analyses of landscape pattern and structure metrics. *Landscape Ecology, 10*(1), 23–39.

Schindler, S., Poirazidis, K., & Wrbka, T. (2008). Towards a core set of landscape metrics for biodiversity assessments: A case study from Dadia National Park, Greece. *Ecological Indicators, 8*(5), 502–514.

Schindler, S., Kati, V., vonWehrden, H., & Wrbka, T. (2009). Landscape metrics as biodiversity indicators for plants, insects and vertebrates at multiple scales. In J. Breuste, M. Kozová, & M. Finka (Eds.), *European landscapes in transformation: Challenges for landscape ecology and management* (pp. 228–231). *European IALE Conference 2009*, Salzburg, Austria & Bratislava, Slovakia.

Skartsi, T., Elorriaga, J., Vasilakis, D., & Poirazidis, K. (2008). Population size, breeding rates and conservation status of Eurasian black vulture in the Dadia National Park, Thrace, NE Greece. *Journal of Natural History*, *42*(5–8), 345–353.

Torras, O., Gil-Tena, A., & Saura, S. (2008). How does forest landscape structure explain tree species richness in a Mediterranean context? *Biodiversity and Conservation*, *17*(5), 1227–1240.

Triantakonstantis, D., Kollias, V., & Kalivas, D. (2006). Forest re-growth since 1945 in the Dadia forest nature reserve in northern Greece. *New Forests*, *32*(1), 51–69.

Yamaura, Y., Amano, T., & Katoh, K. (2008). Relative importance of the area and shape of patches to the diversity of multiple taxa. *Conservation Biology*, *22*(6), 1513–1522.

Chapter 14
Relief for Models of Natural Phenomena

Jana Svobodová and Vít Voženílek

14.1 Digital Representation of Relief

Computer models of spatial phenomena mostly use Digital Elevation Models (DEM) (morphometric characteristics, drainage situation, etc.) to represent the earth's surface and its properties. The definition of the term digital model of relief is currently inconsistent both in Czech and in foreign professional literature (e.g. Šíma, 2002; Voženílek, 2001; Klingseisen, 2004; Longley, Goodchild, Maguire, & Rhind, 2001; Wood, 1996; Dikau, 1989). Burrough & McDonnell (1998) offer an apt definition; for them it is any type of representation of the earth's surface continuously changing in space. In technical practice the term Digital Terrain Model (DTM) is used; however, this term is stricter in its sense (Voženílek et al., 2001). In English written literature the term Digital Elevation Model (DEM; Longley et al., 2001) was introduced, and the abbreviation DEM is now used worldwide. To maintain terminological unity we will use the abbreviation DEM throughout the text.

DEM represents a significant tool in applications processing the earth's surface and is used in geomorphology, hydrology, cartography, climatology, geology and ecology. In the last 20 years many studies have been written in the field of digital relief modelling contributing a lot of information about the earth's surface. Basic types of DEM are commonly used, namely raster grid and vector TIN (triangulated irregular network). The issue of DEMs is elaborated in detail (Krcho, 1990; Voženílek, 1996; Wood, 1999; Rapant, 1996; Wood, 1996). However, the difference in 2D expression of relief on maps that has been used up to now and 3D modelling in DEMs has not been resolved.

The earth's surface is one of the many existing continuous surfaces. The surface can express any continuous spatial phenomenon (air temperature, concentration of substances in the atmosphere, etc.) if the quantity of the given phenomenon is

J. Svobodová (✉)
Department of Geoinformatics, Palacky University in Olomouc, tr. Svobody 26 , 771 46 Olomouc, Czech Republic
e-mail: svobodova-j@centrum.cz

J. Anděl et al. (eds.), *Landscape Modelling*, Urban and Landscape Perspectives 8, DOI 10.1007/978-90-481-3052-8_14, © Springer Science+Business Media B.V. 2010

considered a value z in a 3D coordinate system (Burrough & McDonnell, 1998). Such surfaces are called statistical, they are relatively continuous and can be processed, analysed and represented in the same way as the relief of the earth's surface. In case of the earth's surface the altitude corresponds to the value z.

14.2 Accuracy in Relief Expression

The aim of DEM is to represent the variability of relief as best as possible. However, the quality of such representation is limited by many factors arising from the character of the creation of the model. The load of inaccuracies in the model then influences all analyses performed for this model.

Except for the newest DEMs derived from stereo pairs of aerial or radar photos (using the IFSAR technology, for more see GEODIS, 2008) the majority of DEMs in the Czech Republic are created from digitalised analogue topographical maps of large scales. These maps may not only be outdated but they are also loaded with errors. Their digitalisation augments the errors. We have to keep in mind that it is not possible to perform accurate analyses using inaccurate data and it is essential to maintain a critical approach to the assessment of the results (Voženílek et al., 2001).

There are several sources of errors and inaccuracies in the expression of spatial (3D) properties of relief.

Type of source data: The type of source data influences the accuracy of relief expression. The most common source data are contour lines or height fields (grids of points with their elevation). Both of these ways represent a discrete expression of a continuous relief. Therefore, it is true that the higher the number of accurate discrete objects (contour lines, height points) and the better their distribution (according to the configuration of the relief), the less inaccuracies in relief expression, and thus, in the computer models.

Scale: Thanks to cartographic generalisation the smaller the map scale, the simpler and more generalised the map content, including the expression of altimetry (contour lines, elevation points). When the scale changes, there are changes in the basic contour line interval, contour line course and curvature, as well as the density of elevation points. On a map with a smaller scale a relief expressed by contour lines of 10 m is more smoothed down than a relief with contour lines of 5 m. Except for different expression of detailed surface shapes compared to the real relief a modelled surface has different parameters (area, slope, slope length, etc.). Generally it is true that the smaller the scale, the bigger the inaccuracies in relief expression and the higher the influence on the results of computer modelling. A more exact assessment (quantification) of this dependency will be the subject of further research by the author.

Number of dimensions in the model: The basic influence of the accuracy of relief expression is the change from real 3D to 2D relief expression of maps that most often serve for derivation of values representing the relief in computer

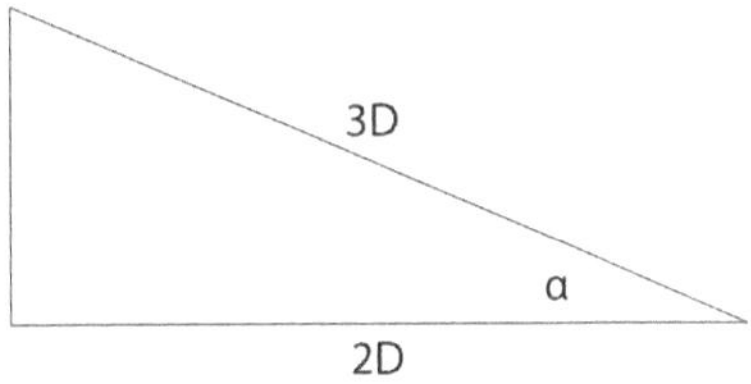

Fig. 14.1 Base of inaccuracies arising from a different 2D–3D relief expression (source: the authors)

models. A map is the most frequent source of altimetric relationships used for determining values for computer models. Projection of the earth's surface to a topographic
projection plane causes the biggest deviations in relief expression. In the case where the region is used as one of the input quantities, a difference in 2D and 3D area appears (Fig. 14.1).

Selection of type of digital elevation model: When generating a DEM the selection of type (grid, TIN) entails another deviation from the reality – the modelled earth surface is represented either by a system of blocks (raster grid type) or an irregular polyhedron (vector TIN type). It is generally true that the TIN type is a much more accurate relief expression than the grid type; nevertheless its structure does not necessarily have to be suitable as the input for all models of physical geographic phenomena.

Selection of interpolation method of DEM: Even when the most suitable altimetric data source is selected, a DEM of the highest accuracy is not necessarily ensured because to generate the model it is necessary to choose a suitable interpolation method, and set its parameters. The selection of an interpolation method and its parameters depends primarily on the relief type. The dependencies of setting the parameters of interpolation methods and the relief types are discussed in the work by Kadlčíková and Tuček (2008).

The generated DEM is used for deriving values of various morphometric and other characteristics of relief (Voženílek et al., 2001); however, their values differ according to the parameters of interpolation methods used. An improperly selected and set interpolation method results in the creation of a DEM of low quality, which then results in the derivation of erroneous values of geomorphometric parameters. Thanks to DEM visualisation, the following Fig. 14.2 shows changes in the course of the relief (and derived morphometric characteristics) depending on the changing values of input parameters of interpolation methods.

Errors in derived parameters are usually much more evident than in the original DEMs. This is further enhanced by the properties (configuration) of the real relief – flat land, hilly areas, highlands and mountainous areas. It is evident that there is direct proportion between the relative relief segmentation and the examined inaccuracies – in hilly areas the inaccuracies are smaller than in mountainous areas (see previous text in this chapter). Any error in the DEM then generates an error in the application results where relief is one of the factors.

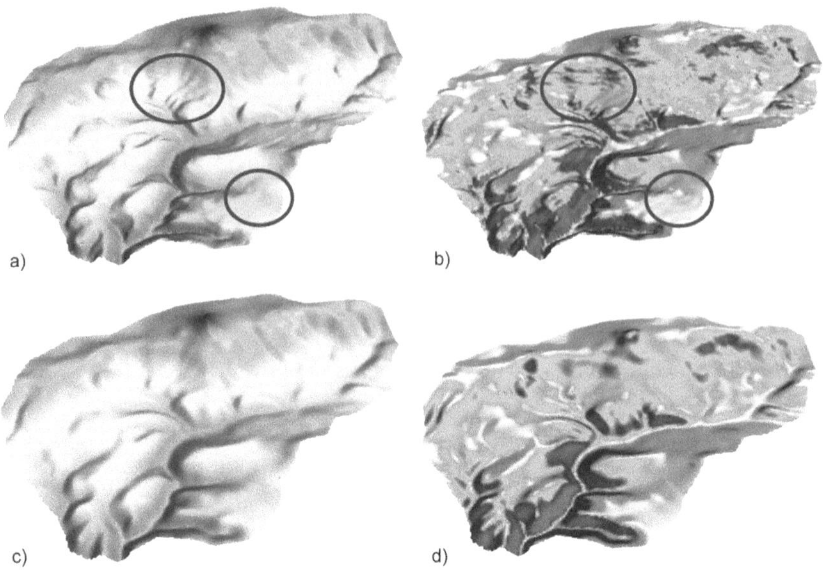

Fig. 14.2 DEMs created using the different settings of spline with tension method (a, c) and grids of slopes derived from these DEMs (b, d)

14.2.1 Methods of Assessment of DEM Quality

The quality of the resulting DEM can be influenced primarily by a suitable selection of the type of source data, scale, relief expression (2D or 3D), DEM type and interpolation method.

There are several methods that can be used for determining the inaccuracies in the DEM. A quicker but less accurate way to detect all errors is a visual check of the DEM (Fig. 14.3a, b). By this we can identify clearly visible traces in case of insufficient vertical resolution and problems with terrain edges or local anomalies (Gallant & Wilson, 2000). However, there are more objective methods comparing DEMs containing assumed errors with reference data using statistical methods. Among the basic means of comparison are standard deviation, average, minimum and maximum values; the more advanced ones include RMSE (root mean square error) calculation, hammock plot or calculation of hammock index. These methods can be used especially for the grid-type DEMs, which thanks to their structure enter the process of modelling various physical geographic phenomena much more often.

The most commonly used degree of inaccuracy is the *root* mean square error – RMSE. It measures the spread of division of the frequency of deviations between the original altimetric data and the DEM data. Mathematically it is expressed in the following way (Wechsler, 1999):

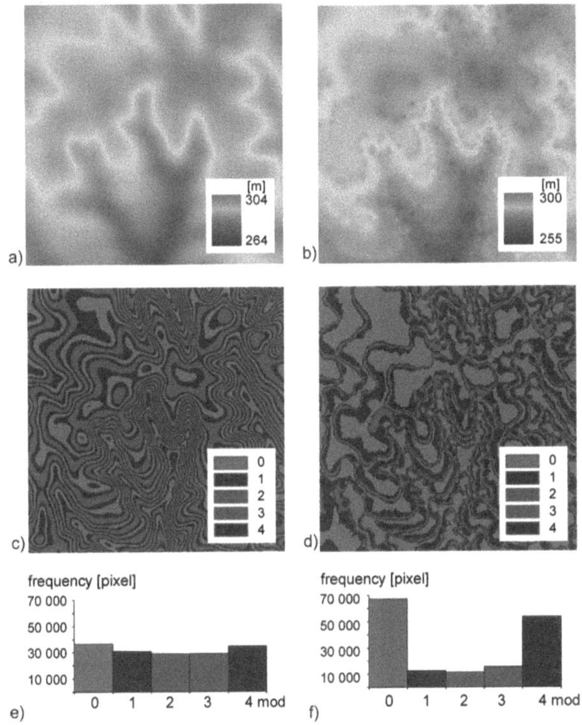

Fig. 14.3 Examples of two DEMs (area of Orlovská plošina) with different properties and examples of hammock plots created from previous DEMs (Orlovská plošina)[2]

$$RMSE_Z = \sqrt{\frac{1}{n}\sum_{i=1}^{n}(Z_{di} - Z_{ri})^2} \tag{14.1}$$

where Z_{di} is the i-th value of altitude from the DEM surface, Z_{ri} is the corresponding original altitude, and n is the number of checked points.

A higher RMSE value means a higher spread between two data sets; an ideal value should not go over half of the value of the interval of the original contour lines. The main advantage of RMSE is the simplicity of the calculation and its clear concept. However, this index represents a simple global degree of deviations, and therefore is not able to fully clarify spatial changes of errors over the interpolated surface. To better understand and quantify DEM inaccuracies, surfaces of the inaccuracies are used more often, especially the hammock plot (primarily when the input data are based on contour lines).

The *Hammock plot* is the surface that originates by whole-number division of values of DEM, the divisor being the value of the basic interval of the input contour lines. Thus, it is possible to detect errors arising from interpolation from contour lines (or possibly points generated from contour lines), which means excessive occurrence of pixel values corresponding to the values of input contour lines – i.e. pixels with the remainder zero. The Hammock plot (i.e. the frequency of particular modules) can also be represented as a histogram.

In Fig. 14.3c, d the red colour represents the areas of module 0, i.e. areas where the values of elevation after rounding off to whole numbers correspond to the values of the input contour lines. The higher the percentage of red colour, the higher influence of values of contour lines during the interpolation and creation of the DEM (Fig. 14.3d). The result of a high-quality interpolation should be a balanced frequency of pixels in all modules (Fig. 14.3c) which represents a better quality (more even) calculation of values of DEM pixels even in areas more remote from the input contour lines.

The Hammock plot is the basis for the calculation of the *hammock index* (H), another characteristic of DEM quality which points out the balance of modules. Its value falls into the interval <–1, i – 1>. Ideal values representing a balanced module (i.e. balanced frequency in all intervals) tend to stay around zero. The formula for the calculation of the hammock index is (Wood, 1996):

$$H = \frac{(nf_0) - \sum_{i=1}^{n-1} f_i}{\sum_{i=0}^{n-1} f_i} \tag{14.2}$$

where n is contour line interval, f_0 is frequency of mod0 (frequency of pixels with the remainder 0), and f_i is frequency of other mod.

The resulting value of the hammock index based on the hammock plot in Fig. 14.3c equals the value 0.48, which means balance of all modules (Fig. 14.3e). However, the value of the hammock index based on the hammock plot in Fig. 14.3d equals the value 2.29, which in the given scope <–1, 4> means a considerable imbalance of modules (Fig. 14.3f) and points out a considerable influence of the values of the original contour lines.

14.2.2 Analysis of Parameters of Interpolation Methods Used for the Creation of DEM in the Czech Republic

On the basis of the above-mentioned assessment it is possible to determine the most suitable interpolation method and its parameters for each type of relief. The following recommendations are based on several works by the author (Kadlčíková, 2007b; Kadlčíková & Tuček, 2008) in which contour lines with the interval 5 m from the digital land model DMÚ 25 (i.e. of the corresponding scale 1:25.000) were used as the input data for testing. The output of the testing was represented by raster grids of individual types of relief, the pixel size being 5 m. The testing regions (Table 14.1) were selected so that they represent all basic relief types on the level of districts as stipulated in the Regional geomorphologic segmentation of relief in

Table 14.1 Districts of geomorphologic segmentation of relief according to type of relief and area (source: the authors)

		Testing regions		Other	
Relief type	RAS [m]	Český masiv– Czech massif	Karpaty– Carpathians	Český masiv– Czech massif	Karpaty– Carpathians
Depression – furrow		Blanická brázda	Pozlovická brázda	Borkovická pánev	
Plain	0–30	Sadská rovina	Žerotínská rovina		Červenecká rovina
Flat hilly land	30–75	Nechanická tabule	Orlovská plošina	Ostroměřská tabule	
Hilly land	75–150	Podještědská pahorkatina	Vlčnovská pahorkatina	Radomyšlská pahorkatina, Milínská pahorkatina	Prakšická pahorkatina
Flat highland	150–220	Studenská vrchovina	Divácká vrchovina		Uhřická vrchovina
Highland	225–300	Bozkovská vrchovina	Hošťálkovská vrchovina	Kozlovská vrchovina, Rožmberská vrchovina	
Flat mountains	300–450	Ústecké středohoří	Rusavská hornatina		
Mountains	450–600	Hornoopavská hornatina, Černohorská rozsocha	–	Boubínský hřbet, Včelenská hornatina	

RAS: relative altitudinal segmentation.

the Czech Republic both from the area of Český masív and Karpaty. The reason for selecting data from both areas is their different geological evolution and the land's character.

Primary testing was performed on 16 testing regions. Table 14.2 shows that the spline with tension method, which creates surfaces of various degree of smoothness according to the set weight value, is suitable for all types of relief.

The essence of this method makes it suitable for the interpolation of relief using altimetric data (ESRI ArcGIS 9.2 desktop help). This supposition was confirmed during the testing of use of this and other methods for calculating various relief types. The values of the parameter number of input points are most commonly 20–30 points for all relief types, higher values of this parameter led to erroneous interpolation due to the fact that very remote points were also included. For less segmented relief types a lower number of input points is more suitable because with values around 30 it can happen that more remote points are included due to their low density. For the parameter of weight the spread of values is from 1 to 10 for all relief types. Therefore, for relief modelling it is suitable to use lower values of the parameter weight of the spline with tension method.

Table 14.2 The most suitable method and setting of its parameters to achieve the best quality DEM for individual relief types (source: the authors)

	Karpaty – Carpathians			Český Masiv – Czech massif		
Relief type	Interpolation method	No. of input points	Weight	Interpolation method	No. of input points	Weight
Plain	Spline with tension	10	10	Spline with tension	5	5–10
Flat hilly land	Spline with tension	20	1	Spline with tension	10	1–5
Hilly land	Spline with tension	20	1	Spline with tension	20	1
Flat highland	Spline with tension	30	10	Spline with tension	30	10
Highland	Spline with tension	20	1–5	Spline with tension	30	1–5
Flat mountains	Spline with tension	30	1–5	Spline with tension	30	5
Mountains	–	–	–	Spline with tension	15	1–5
Furrow	Spline with tension	20	1	Spline with tension	20–30	1–5

Even though the results are contributive, for more objective results it is necessary to continue testing, extend it with other areas, compare the results and observe whether there is a trend.

14.2.3 Influence of DEM Quality on 3D Relief Expression

Thanks to projection of the earth's surface to the topographic projection plane there arise deviations both in the relief expression and in its geometrical characteristics. In case the area of a region or the slope length is used as one of the input quantities during modelling physical geographic phenomena, we must keep in mind that using data acquired by 2D measurement is a mistake.

The question is whether the mistake is important for us or not. Table 14.3 shows that for small values of slope, which can also correspond to low segmentation of terrain, the deviations measured in 2D and 3D are negligible. However, with increasing slope the value measured in 3D can be up to double; in that case the results of models of physical geographic phenomena can be influenced more considerably.

The higher the segmentation of the terrain, the bigger the errors. Firstly, there are the above-mentioned differences in 2D and 3D expression. Secondly, there are errors in the quality of the DEM which influences the derived parameters.

When comparing deviations of basic statistical characteristics from input altimetric data for example in highlands and flat land, it is clear that deviations (errors) and differences between the results of individual interpolation methods decrease from more segmented relief types to less segmented relief types (Tables 14.4 and 14.5).

Table 14.3 Differences of lengths and areas in DEM relief expression compared to expression in maps (source: Voženílek, 2002b)

Slope (°)	Length 100 m along fall line [m]	Area 100 × 100 m [1 ha]
2	100.06	1.0006
5	100.38	1.0038
10	101.54	1.0154
15	103.52	1.0352
25	110.34	1.1034
35	122.08	1.2208
55	174.34	1.7434

Table 14.4 Deviations of basic statistical characteristics from the values of input points for tested methods; example of flat highland (source: the authors)

Divácká vrchovina (flat highland)		Average altitude [m]	Standard deviation [m]	Max. altitude [m]	Min. altitude [m]
Input points		284.060	29.340	220.000	385.000
		Deviation from input points [m]			
IDW	Max.	0.812	2.802	7.855	16.972
	Min.	0.537	0.456	0.000	10.001
Regulated spline	Max.	1.034	3.499	10.562	9.113
	Min.	0.967	3.302	3.165	8.848
Spline with tension	Max.	0.964	3.319	2.343	9.528
	Min.	0.795	2.724	0.148	9.060

Table 14.5 Deviations of basic statistical characteristics from the values of input points for tested methods; example of plain (source: the authors)

Žerotínská rovina (plain)		Average altitude [m]	Standard deviation [m]	Max. altitude [m]	Min. altitude [m]
Input points		251.713	3.840	245.000	260.000
		Deviation from input points [m]			
IDW (Inverse Distance Weighting)	Max.	0.825	1.768	1.919	3.477
	Min.	0.604	0.128	0.000	0.000
Regulated spline	Max.	0.877	0.400	0.149	0.585
	Min.	0.802	0.275	0.000	0.000
Spline with tension	Max.	0.880	0.411	2.966	3.086
	Min.	0.630	0.196	0.000	0.000

Table 14.6 Maximum and minimum RMSE values acquired by using individual interpolation methods; example of flat highland (source: the authors)

Divácká vrchovina (flat highland)					
			Setting of parameters		
RMSE		Interpolation method	No. of input points	Power	Weight
Max.	6.078	IDW	50	0.5	
Min.	2.096		10	3.0	
Max.	0.820	Regulated spline	10		0.9
Min.	0.722		30		0.0
Max.	1.142	Spline with tension	3		50.0
Min.	0.719		30		10.0

Table 14.7 Maximum and minimum RMSE values acquired by using individual interpolation methods; example of plain (source: the authors)

Žerotínská rovina (plain)					
			Setting of parameters		
RMSE		Interpolation method	No. of input points	Power	Weight
Max.	1.824	IDW	50	0.5	
Min.	0.000		3	2.0	
Max.	0.153	Regulated spline	5		0.00
Min.	0.110		10		0.01
Max.	0.257	Spline with tension	3		50.00
Min.	0.097		10		10.00

This is conditioned by decreasing spread of values of elevation. A similar trend can be seen in the RMSE. Differences in the RMSE values both between the tested methods and within individual methods decrease from more segmented relief types (mountainous areas, highlands) to less segmented relief types (hilly areas, flat lands; Tables 14.6 and 14.7).

The influence of the DEM quality on the derived geometrical parameters in 3D is shown in Table 14.8. As we have already mentioned errors in more even relief are minimum, and therefore, the difference in the area measured in 3D in a higher-quality and lower quality DEM is negligible. However, in highlands or mountainous areas the difference between the area measured in a higher quality and lower quality DEM is more considerable, which could have some influence on further modelling of physical geographic phenomena.

Table 14.8 Changes of parameters of modelled relief in different relief types of the scale 1:25,000 with the basic contour line interval 5 m (source: the authors)[3]

				High-quality DEM			Low-quality DEM		
Relief type	RAS [m]	LCL [km]	2D area [km^2]	3D area [km^2]	W*	NOP*	3D area [km^2]	W*	NOP*
Flat land	0–30	9.370	4	4.000	10	15	4.004	0	20
Hilly area	30–150	103.258	4	4.051	1	20	4.070	0	20
Highland	150–300	190.690	4	4.158	5	30	4.298	0	20
Mountain area	300–600	278.485	4	4.288	5	15	4.576	0	20

14.3 Terrain Sensitivity

Terrain sensitivity is the ability of digital relief expression to influence the results of spatial modelling, which includes influence by the earth's surface. Terrain sensitivity in models of spatial phenomena depend on all stages of surface processing, starting with the selection of source data, through to determination of resolution, interpolation methods and their parameters, etc. It is caused by generalised expression of the earth's surface in digital data structures (Voženílek, 2002a).

Physical geographic models also include various derived parameters of the earth's surface, e.g. morphometric characteristics and shape of relief. This is the reason why a number of studies (e.g. Etzelmuller, 2000; Wolock & McCabe, 2000; Tucker, Catani, Rinaldo, & Bras, 2001) discuss the suitability of using source data for relief expression. However, they take into account generalised sources of altimetric data (contour lines) whereas to express relief more accurately and to decrease the influence of inaccurately expressed relief it is necessary to use hypsometric representation as close to primary sources of altimetric data as possible (Schoorl, Sonneveld, & Veldkamp, 2000).

14.4 Influence on Modelling Physical Geographic Phenomena, Using the Example of Erosion Processes

Development of modelling of erosion processes currently clearly heads towards mathematical expression of subprocesses leading to soil erosion and towards implementing such a model in the GIS environment. From the point of view of acquiring input information the GIS is becoming an essential source. In the area of acquiring data about relief (inclination, orientation, slope length, curvature and drained areas) DEMs represent a dominant development environment. The speed and accuracy of calculation is crucial. Remote sensing is a source of further information, concerning especially land use and seasonal changes in soil qualities. Taking into account the specificity of some data it is necessary to acquire them by field research, or potentially by laboratory work.

Modelling of erosion processes most often includes creating mathematically-expressed models that make use of knowledge about soil erosion as a natural phenomenon. Therefore, there are models of various types (Voženílek, 1999):

- deterministic models based on analytical-physical knowledge;
- conceptual models emphasizing the impact of topography on erosion and accumulation of material;
- empirical-statistical models based on laboratory methods of determining values and on the importance of input factors.

The connection of some models to GIS technologies enables them to become part of more comprehensive optimisation models (methodology), especially in relation to spatial optimisation of land structure. The efficiency of these models can also enhance their interconnection (within GIS) with models of other geomorphologic processes which interact with water erosion (creep, etc.). A number of models (e.g. CREAMS, ANSVERS, SMODERP, etc.) enable modelling of soil erosion on individual slopes or in very small river basins. In these models the impact of relief representation on the modelling results can be seen via the models' factors.

14.5 Conclusions

The issue of relief expression for the purposes of modelling physical geographic phenomena of land deserves constant attention, especially because of the impact of different approaches when incorporating quantities related to relief, which is most often the creation and use of digital relief models. Inaccuracies that arise during the process of relief modelling are reflected further in the modelling results. Even when using one set of source maps DEMs of various qualities can be generated due to different input parameters. These then lead to deviations of various significance when used in computer models of physical geographic phenomena.

The extent of errors increases primarily with the increasing segmentation of relief. Firstly, there are errors (differences) arising from expression in 2D or 3D. Secondly, there are errors in the quality of the DEM which influences the derived parameters. Low segmentation of relief, to which also small slope inclination of the land can correspond, leads to negligible deviations measured in 2D and 3D. However, with increasing inclination the value measured in 3D can be up to double; in that case the results of models of physical geographic phenomena are influenced more considerably. When assessing DEM quality using basic statistical characteristics or RMSE it is also possible to observe the following trend: errors (differences) arising from the use of different interpolation methods (and different setting of their parameters) compared to reference data decrease from more segmented to less segmented relief types, which is conditioned by the decreasing spread of values of elevation.

Notes

1. The DEM created using the spline with tension method with the parameters tension = 40, smooth = 0.1(a); derived from this a grid of the slopes (b) where there are marked selected places with undesirable surface unevenness arising due to erroneous interpolation. DEM created using the spline with tension method with the parameters tension = 20, smooth = 0.8 (c); derived from this a grid of the slopes (d) where there are no errors visible "at first sight". However, for the final determination of the most suitable interpolation parameters it is always necessary to perform statistical evaluation (e.g. an RMSE calculation) (source: Kadlčíková, 2007a)
2. For DEMs: it is possible to visually identify (a) a minimum (spline with tension method, weight 15, no. of points 30) and (b) a high (IDW method, power 2, no. of points 50) number of gross errors (blunders). Hammock plots representing individual modules after division by 5 – red mod0 (i.e. the remainder after the division is 0), violet mod1 and mod4, blue mod2 and mod3 in the form of surfaces (c–d) and in the form of histograms (e–f) (source: the authors)
3. Grid DEM modelled by the spline with tension method. Setting for high-quality DEM is based on Table 14.2; RAS: relative altitudinal segmentation; LCL: length of contour lines; W: weight; NOP: number of points. (source: the authors)

References

Burrough, P. A., & McDonnell, R. A. (1998). *Principles of geographical information systems*. Oxford: Oxford University Press.

Dikau, R. (1989). The application of a digital relief model to landform analysis in geomorphology. In J. Haper (Ed.), *Three dimensional application in geographic information systems* (pp. 51–77). London: Taylor & Francis.

ESRI. *ArcGIS 9.2 Desktop Help*, http://webhelp.esri.com/arcgisdesktop/9.2/

Etzelmuller, B. (2000). On the quantification of surface changes using grid-based digital elevation models (DEMs). *Transactions in GIS, 2*, 129–143.

Gallant, J. P., & Wilson, J. C. (2000). *Terrain analysis – Principles and applications*. New York: John Wiley & Sons.

GEODIS. (2008). *Radar map – Intermap*, http://www.geodis.cz/sluzby/radarova-mapa-intermap

Kadlčíková, J. (2007a). *DMR a Jeho Vizualizace, Proceedings of the Conference GIS Ostrava 2007*, TU-VSB, Ostrava (CD-ROM).

Kadlčíková, J. (2007b). Testování a výběr interpolačních metod DMR v závislosti na typu georeliéfu. *Geomorphologia Slovaca et Bohemia, 7*(2), 14–18.

Kadlčíková, J., & Tuček, P. (2008). *Evaluation and setting of parameters in Interpolating Methods by Modeling of Different Types of Georelief, Proceedings of the Conference GIS Ostrava 2008*, TU-VSB, Ostrava, p. 8.

Klingseisen, B. (2004). *GIS Based Generation of Topographic Attributes for Landform Classification*, Diploma thesis, Fachhochschule Technikum, Kärnten.

Krcho, J. (1990). *Morfometrická Analýza a Digitálne Modely Georeliéfu*. Bratislava: Veda.

Longley, P. A., Goodchild, M. F., Maguire, D. J., & Rhind, D. W. (2001). *Geographic information systems and science*. New York: John Wiley & Sons.

Rapant, P. (1996). *Digitální modely terénu*. PGS, TU-VSB, Ostrava.

Schoorl, J. M., Sonneveld, M. P. W., & Veldkamp, A. (2000). Three-dimensional landscape process modelling: The effect of DEM resolution. *Earth Surface Processes and Landforms, 25*(9), 1025–1034.

Šíma, J.(2002). *Musíme používat pracovní slang při prezentacích a v publikacích o geografických informačních systémech? Proceedings of the Conference GIS Ostrava 2002*, TU-VSB, Ostrava (CD-ROM).

Tucker, G. E., Catani, F., Rinaldo, A., & Bras, R. L. (2001). Statistical analysis of drainage density from digital terrain data. *Geomorphology, 36*(3), 187–202.

Voženílek, V. (1996). *Fundament of digital elevation model as a tool for geomorphological research*. Olomouc: Acta Univ. Palacki. Olomouc – Geographica 34, Palacky University.

Voženílek, V. (1999). Geoinformační aspekty modelování eroze půdy. In V. Voženílek (Ed.), *Integrace prostorových dat, proceedings of the conference "Integration of Spatial Data"* (pp. 208–232). Olomouc: Palacký University.

Voženílek, V. (2002a). Terrain sensitivity in environmental models. *Geografie – Sborník ČGS, 107*(2), 111–120.

Voženílek, V. (2002b). *Vliv vyjádření reliéfu na výsledky modelování prostorových jevů, Proceeding of the Sloval Geographical Society Congress*, UKF, Nitra.

Voženílek, V., Kirchner, K., Konečný, M., Kubíček, P., Létal, A., Petrová, A., et al. (2001). *Integrace GPS/GIS v Geomorfologickém Výzkumu*. Olomouc: Palacky University.

Wechsler, S. P. (1999). *Digital Elevation Model (DEM) Uncertainty Evaluation and Effect on Topographic Parameters, ESRI User Conference*. ESRI, Redlands.

Wolock, D. M., & McCabe, G. J. (2000). Differences in topographic characteristics computed from 100- and 1000-m resolution digital elevation model data. *Hydrological Processes, 14*(6), 987–1002.

Wood, J. D. (1996). *The Geomorphological Characterisation of Digital Elevation Models*, PhD Thesis, Geography Department, University of Leicester.

Wood, J. D. (1999). Modelling the continuity of surface form using digital elevation models. In T. Poiker & N. Chrisman (Eds.), *Proceedings, 8th international symposium on spatial data handling* (pp. 725–736). Vancouver: Simon Fraser University.

Name Index

DOI 10.1007/978-90-481-3052-8,

Subject Index

FSC
www.fsc.org
MIX
Papier | Fördert
gute Waldnutzung
FSC® C083411

Kolibri 360 GmbH
Ferdinand-Jühlke-Straße 7
99095 Erfurt, Deutschland
produktsicherheit@kolibri360.de